The Military's Business

If the military were a business, would you buy shares? Over recent years, Western armed forces, particularly the US, have been costing more yet achieving less. At the same time, austerity measures are reducing defence budgets. This book uses defence data to examine the workings of modern Western militaries and explore what kind of strategies can overcome this gap between input and output. Instead of focusing on military strategy, Mikkel Vedby Rasmussen seeks to draw on the ideas of business strategy to assess alternative business cases – reforming military HR to combat instability in the 'Global South' or utilising new technologies to overcome the prohibitive costs of current systems. Analysing the philosophical, strategic and budgetary underpinnings of these alternatives, he concludes that a more radical break from current military organisational practices is needed which would allow them to fit within a nation's overall national security system without ever-increasing budgets.

MIKKEL VEDBY RASMUSSEN is a professor at the University of Copenhagen. He is currently on leave working on policy development in the Danish Ministry of Defence. Rasmussen has previously headed the Centre for Military Studies and the Danish Institute for Military Studies in Copenhagen. He has published widely on security and defence matters, including *The Risk Society at War* (Cambridge University Press, 2006).

The Military's Business

Designing Military Power for the Future

MIKKEL VEDBY RASMUSSEN

CAMBRIDGE
UNIVERSITY PRESS

University Printing House, Cambridge CB2 8BS, United Kingdom

Cambridge University Press is part of the University of Cambridge.

It furthers the University's mission by disseminating knowledge in the pursuit of
education, learning and research at the highest international levels of excellence.

www.cambridge.org
Information on this title: www.cambridge.org/9781107477353

© Mikkel Vedby Rasmussen 2015

This publication is in copyright. Subject to statutory exception
and to the provisions of relevant collective licensing agreements,
no reproduction of any part may take place without the written
permission of Cambridge University Press.

First published 2015

Printed in the United Kingdom by Clays, St Ives plc

A catalogue record for this publication is available from the British Library

Library of Congress Cataloguing in Publication data
Rasmussen, Mikkel Vedby, 1973–
The military's business : designing military power for the future /
Mikkel Vedby Rasmussen.
    pages   cm
Includes bibliographical references and index.
ISBN 978-1-107-09477-2 (hardback) – ISBN 978-1-107-47735-3 (pbk.)
1. Military art and science–United States–History–21st century.   2. United
States–Armed Forces.   3. War–Forecasting.   I. Title.
U43.U4R48 2015
355.020973–dc23
2014043074

ISBN 978-1-107-09477-2 Hardback
ISBN 978-1-107-47735-3 Paperback

Cambridge University Press has no responsibility for the persistence or accuracy of
URLs for external or third-party internet websites referred to in this publication,
and does not guarantee that any content on such websites is, or will remain,
accurate or appropriate.

Contents

Figures

Acknowledgements

Since 2006 I have been involved in strategic innovation as the director of a government-funded research centre – first at the Danish Institute for Military Studies and then at the Centre for Military Studies at the University of Copenhagen. This has given me ample opportunity to engage with the military and military strategy in Denmark, the Nordic countries and on the European level. I have increasingly found that the way military strategy is formulated and the way in which military forces are structured to implement this strategy is taken for granted in ways that in fact make military strategies fail and structure military forces in wasteful ways. It is my impression that there is a widespread understanding of this and there have been numerous attempts to reform elements of the military business, but many shy away from asking the obvious question: are there alternative ways of doing the military business? Thus, I became interested in how one would conceptualise new ways of doing strategy. This interest has been furthered by the fact that I also teach strategy at various military educational establishments, including the NATO Defense College, as well as at the University of Copenhagen. Explaining the state of strategy to my students, I realised, required a more profound understanding of what the military was doing and how this translated to strategy. This pedagogical project is equally important when discussing with politicians and generals what the future priorities of the armed forces should be. The years of austerity have given ample opportunity to do so and this underlined the need to take a step back and look at the military system as well as the strategies it produces. This book is the result of my reflections and

I hope it can contribute to the reassessment of the foundations of military strategy which are taking place at the moment.

I would like to thank my colleagues at the Centre for Military Studies. One of the hallmarks of a good academic working environment is that the stuff you casually discuss at lunch ends up inspiring you to write a few pages after lunch – and I think my colleagues will find that this has also been the case in this book. Karen Lund Pedersen was kind enough to read an early version of Chapter 2. While I was writing this, Karen and I have set up a master's programme in strategy and the discussions about this have been very instructive for my work on the book. So has the teaching of a summer school on Danish defence policy with my colleague Henrik Breitenbauch. Thanks go also to the Centre for Military Studies research assistants Jacob Petersen, Mads Boss, Lasse Poulsen, Casper Sakstrup and Mikkel Broen Jakobsen. Thanks also to Jon Jay Neufeld for correcting at least some of the mistakes of a non-native English speaker. Special thanks goes to the two anonymous reviewers at Cambridge, who gave me a lot to work with, and plenty of time to do it, as well as to my editor, John Haslam.

Tradition has it that at this point I should thank my wife for keeping children and various domestic animals away while I was writing, but, though Anne in fact gave me that space, her real contribution has been good spirits and a creative mind which keeps me inspired.

The effort these people have put into helping this work along only underlines the fact that any errors and omissions are all my responsibility.

1   **Introduction**

> I'll tell you this much. War creates a closed world and not only for those in combat but for the plotters, the strategists. Except their war is acronyms, projections, contingencies, methodologies … They become paralyzed by the system of their disposal.
>
> Don Delillo[1]

In January 2012, two helicopters landed discreetly on the outskirts of a Somali coastal village and a Navy SEAL team disembarked the helicopters, making their way to the village. There, Somali 'pirates' held an American and a Dane hostage. After a brief firefight, eight of the hostage-takers lay dead and the SEAL team withdrew with the hostages.[2] This operation is a simple example of the decisiveness of military force. There are few other tools at the disposal of a government that produce results that are as immediate and unequivocal as the use of military power. This decisiveness can be deceptive, however. The daring SEAL team rescue neither put an end to Somali lawlessness nor addressed the causes of piracy in the area. And the Navy SEAL team would soon have to go on another mission. Such is the paradox of military power: armed force is the most decisive instrument of a state's power, but the nature of military power is such that the decisiveness is as partial as it is ephemeral. The art of war involves using the right measure of force at the right time for the right ends in acknowledgement of the paradox of military power. Thus, the paradox of military power must be upmost in one's mind when evaluating statements about militaries like the one made by US President Barack Obama in September 2012, when he declared that 'the United States military is the finest fighting force

---

[1] Don Delillo, *Point Omega* (London: Picador, 2010), 35.

[2] 'Raid In Somalia: U.S. Navy SEALs Free Hostages Held Since October', *Huffington Post*, January 2012, www.huffingtonpost.com/2012/01/25/raid-in-somalia_n_1230062.html (2 February 2012).

in the history of the world'.[3] Even if that statement might upset the ghost of a German or Roman general, the US military is incredibly accomplished in combined arms operations, but are such skills the ones most required to ensure the safety of America and the free world? And are the platforms necessary for such operations something the Western taxpayers can continue to pay for? President Obama's statement raises more questions than it answers; and this is the case of Western defence policy in general in the first decades of the twenty-first century. Western defence budgets are increasingly governed by the law of diminishing returns, but generals and policy-makers appear unable to reform the armed forces. The result is that Western military forces are ill-prepared for a century of transformation because they seem incapable of transforming themselves.

Perhaps one reason why Western armed forces appear to be caught in the paradox of military power is that generals and policy-makers seem to take the nature and utility of armed force for granted. For all of the hundreds of books and articles written on strategic challenges and future risks, there are surprisingly few reflections on how the military organisation is able to deal with these challenges and risks. The paradox of military power is rarely recognised in our discourse on military force, which seems to presuppose that the armed forces deliver a certain outcome when and if deployed. What is in fact a vexing paradox is often presented as a straight-forward equation whereby military output is a function of the defence budget. In fact, defence budgets seem to buy less and less.

Such is the paradox explored in this book. Why are Western armed forces achieving less and costing more? What kind of strategies can overcome this disproportion between input and output? This question is often asked and answered with the enemies of the US military and its allies in mind. This is a relevant consideration, which will be made in the course of this book. Given the fact that the

---

[3] Barack Obama, *Remarks by the President and First Lady on the End of the War in Iraq*, The White House, 2011, www.whitehouse.gov/the-press-office/2011/12/14/remarks-president-and-first-lady-end-war-iraq, (30 September 2012), §26.

United States and its allies still enjoy crushing military superiority, however, it might also be prudent to find an answer to the question of why Western armed forces are achieving less and costing more by studying how Western militaries go about their business. This book thus investigates what might be termed the 'military business model'. However, instead of using military strategy, which would be the traditional choice when working to understand the armed forces, I will turn to business strategy.

Military strategy and business strategy approach their respective subjects differently, which is also true of strategic studies and business studies. In the armed forces and among the scholars studying defence and security issues, strategy is about identifying the crucial issues and dealing with them using the right tools from the national security toolbox. A strong premium is placed on realism and common sense, so no great points are given for observing that another type of policy ten years back might have eliminated the need to act now. The fact that the United States probably should not have allowed the Shah to begin a nuclear programme in the 1970s is of little help to the President in determining whether or not to bomb the nuclear facilities at Esfahan and elsewhere in Iran. By the same token, the President's military advisers cannot dream up capabilities for dealing with the nuclear sites that are not in the US inventory (e.g., beam weapons from orbit). As Secretary of Defense Donald Rumsfeld commented in 2004, 'as you know, you go to war with the Army you have'.[4] It is somewhat interesting that Secretary Rumsfeld refers to this as common knowledge, because that would hardly be the approach in the business world where Donald Rumsfeld had success in reorganising a number of firms.

Whereas military strategy deals with the application of an organisation's (the armed forces) resources to deal with a certain problem, business strategy usually focuses on how to frame an organisation to

---

[4] US Department of Defense, *News Transcript: Secretary Rumsfeld Town Hall Meeting in Kuwait*, 8 December 2004, www.defense.gov/transcripts/transcript.aspx?transcriptid=1980 (25 September 2012).

best deal with a certain problem. Where military strategists are supposed to stick to the issue at hand, business strategists are supposed to redefine the issue. In *Game-Changing Strategies*, Constantinos Markides observes how successful business innovation is about redefining the field, redefining the customer and redefining what a business is really offering the customer.[5] Rent-A-Car became the largest car rental company in the United States by redefining the customer from being a traveller arriving at the airport needing a car for local transport to locals in need of a car while their own is at the garage. Rent-A-Car therefore placed their offices downtown rather than at the airport and created a new car rental market.[6] In technical terms, Rent-A-Car redefined the business model for automobile rentals. A business model is the real answer to a few of the questions that are considerably easier to ask than to answer. Constantinos Markides lists the most important of these questions: 'Who should I target as customers? What products or services should I be offering them and what should be my (differentiated) value proposition? How should I do this in an efficient way?'[7]

Because military strategy has a different focus, questions are seldom posed on the 'military business model'. Even though Cicero was the first to note that money is the sinew of war, military strategists have the view of the nobility that money is something you have, not something you talk about. Beatrice Heuser points out that 'those thinking war will always hope for it to be neater, more coherent, purposeful, goal-oriented than it turns out to be in practice'.[8] She describes this more intellectually 'pure' approach to strategy as being in opposition to bureaucratic politics. Heuser does so in an epilogue to a fascinating account of paradigms of military

---

[5] Constantinos C. Markides, *Game-Changing Strategies: How to Create New Market Space in Established Industries by Breaking the Rules* (San Francisco, CA: Jossey-Bass, 2008), 28–46.

[6] *Ibid.*, 35.

[7] *Ibid.*, 6.

[8] Beatrice Heuser, *The Evolution of Strategy: Thinking War from Antiquity to the Present* (Cambridge University Press, 2010), 499.

strategy-making from antiquity to the present day. Heuser's clear conclusion is that the implementation of strategy in staff work, logistical arrangements, fuel and fodder, payment of the troops and so on is an afterthought in the literature on military strategy. This reflects the closed system in which military strategy is discussed and determined but does not reflect the actual implementation of strategy. There is a saying among staff officers that those interested in warfare talk strategy, whereas those that actually know anything about it talk logistics.

The lack of attention to how and by whom strategic concepts are actually implemented is even more curious when considering the fact that these have always been an important aspect of strategic thinking, even if the academic approach to strategy seems to have edited them out. One of the first coherent treatises on military strategy is Sun Tzu's *The Art of War*. The background for its lofty axioms was a very concrete socio-economic transformation of Chinese society that profoundly changed the business model for war. The production of iron weapons, the breeding, feeding and training of a large number of cavalry horses and the troops to use them strained the finances of the Chinese rulers. Where noble warriors in chariots had been specialised in warfare and their campaigns therefore of no great consequence to the rest of society, the levy of peasants required to fight as infantry and the maintenance of cavalry formations not only strained the economy of an agricultural society, it also brought social instability to the countryside that might result in discontent and, in the worst case, rebellion. Discontent might also arise from the expense of provisioning troops and horses. Little wonder that Sun Tzu states that 'those adept in waging war do not require a second levy of conscripts nor more than one provisioning'.[9]

In the twentieth century, Britain used up half of its GDP on warfare in the two world wars. The size of its GDP enabled Britain, with American help, to out-produce and out-fight Germany, but the

[9] Sun Tzu, *The Art of War*, transl. and with an introduction by Samuel B. Griffith (Oxford University Press, 1963), II, 9.

bills for the two wars had a profound impact on British society. The Eisenhower administration wanted to avoid paying such a price for the Cold War confrontation with the Soviet Union and attempted, ultimately unsuccessfully, to rely on nuclear weapons for defence because doing so was less expensive than a large standing army. The predicament of Britain's public finances following the 2008 financial crisis led Paul Cornish to argue that the first duty of the UK government in times of austerity is not the security of its citizens but maintaining a triple-A credit rating;[10] that is a national strategy for maintaining wealth and influence in a time of crisis and globalisation, which means defence cuts. From Sun Tzu to the present, a society's capacity to pay for armed force has shaped force structure and strategy profoundly. As Paul Cornish points out, this is especially pertinent in times of austerity. Many of the subjects discussed in this book stem from this very need to rebalance the books and make defence face its share of the cuts.

Even a cursory glance over military history tells you that the organisation and financing of warfare is an important element in the story – even if behind the scenes. It is not for nothing that military strategists refer to the theatre of war. It is on the battlefield that the real drama unfolds, in the light of which the organisational, logistical and financial aspects that are at the centre of the business strategist's attention seem somewhat pale. Even if the business world seems less colourful than the battlefield, civil society has traditionally held a suspicion that the armed forces were less than efficient in their use of the taxpayers' money. Especially in liberal societies, business practices were viewed as a blueprint for a more rational way of organising the military business. 'No general staff or other system can be made to function properly', US Secretary of War Elihu Root argued in 1901, when 'the present unbusiness-like method prevails of having a Secretary of War and a commanding general of the army

---

[10] Paul Cornish, *Strategy in Austerity: The Security and Defence of the United Kingdom*, 16 October 2010, www.chathamhouse.org/publications/papers/view/109490 (9 September 2014), 1.

to control matters'.[11] Secretary Root tried to introduce Taylor's concepts of scientific management – with no great success.[12] After the Second World War, however, many of Secretary Root's ideas were reintroduced. Mark Grandstaff notes how 'several changes in the personnel policy encouraged military leaders to view themselves as business leaders ... making the military analogous to a large business complete with profit motive, competition, and scientific efficiency, it seemed more corporative, more American, more familiar'.[13] Today, the US Department of Defense's Business Board, which issues reports and guidance to the Secretary of Defense, exemplifies this approach.

But it was not just the military becoming more business-like; there was also a tendency in the business world towards the adoption of the military's idea for making strategies. Strategy is a conceptual practice defined by ideas as well as action.[14] Strategy is only relevant for plotting a course in a particular situation. As business students sometimes realise at their peril, there is little point in making an abstract business model. Strategy is a concrete enterprise, but the point of departure for strategy is also that experience is not enough. As a conceptual tool, strategy is necessary because the general or CEO requiring a strategy cannot develop one without reflecting on the nature of their situation in more general terms. Alfred Sloan's reorganisation of General Motors in 1921 has been referred to as the first self-conscious strategy process.[15] If beginning with the development of strategy as a concept in publishing on business affairs, then Peter Drucker's *Concepts of the Corporation* from 1946 and Alfred Chandler's *Strategy and Structure* from 1962 are academic milestones which can be used to measure the development of

---

[11] Quoted in Donald Vandergriff, *The Path to Victory: America's Army and the Revolution in Human Affairs* (Novato, CA: Presidio Press, 2002), 48.

[12] *Ibid.*, 55.

[13] *Ibid.*, 79.

[14] Anne Swidler, 'Culture in Action: Symbols and Strategies', *American Sociological Review* 51 (1986).

[15] Richard Koch, *Strategy: How to Create, Pursue and Deliver a Winning Strategy* (Harlow: Prentice Hall, 2011), 7.

the concept of strategy.[16] Strategy comes into being in the 1960s as a practice combining ideas about good strategy with concrete action in firms conducting strategy processes.[17] Lockheed was one of the corporations in which strategy was put into practice by Igor Ansoff and others who realised that a defence company needed a business strategy to match the national security strategy.[18] Ansoff published his principles in *Corporate Strategy* in 1965 – a book which became 'the bible of strategic planning'.[19] Lockheed required a strategy for how to adapt its production to national security priorities. To Lockheed, strategy was an organisational matter, whereas the US government perceived it as being a matter of the global threat from the Soviet Union. The government had already let the notion of strategy out of the conceptual bag, however. In NSC68, the US government defined national security as representing much more than the mere military capability to deter Soviet aggression. It was the US ability to maintain a free and prosperous society that would prove the ultimate argument in the ideological confrontation of the Cold War, as well as providing the material foundation for pursuing military deterrence, the National Security Council argued.[20] Business and military strategies possibly began to diverge conceptually even at this early stage. Even if military and corporate strategies pursued a common national agenda, strategy was institutionalised very differently. Military strategy remained an in-house government activity with limited, severely regulated access granted to outsiders. One illustration of this might be the fact that NSC68 was written in 1950 but first declassified in

---

[16] *Ibid.*, 7.

[17] *Ibid.*, 7. Lawrence Freedman, *Strategy: A History* (Oxford University Press, 2013), 491–504.

[18] Yves Doz and Mikko Kosonen, *Fast Strategy: How Strategic Agility Will Help You Stay Ahead of the Game* (Harlow: Wharton School Publishing, 2008), xii.

[19] Koch, *Strategy*, 7.

[20] 'A Report to the National Security Council by the Executive Secretary on United States Objectives and Programs for National Security', 14 April 1950, reprinted in Dennis Merrill (ed.), *Documentary History of the Truman Presidency, Volume 7: The Ideological Foundations of the Cold War – the 'Long Telegram', the Clifford Report, and NSC 68* (University Publications of America, 1996).

1975. The relationship between those conducting strategy and those studying it could only occasionally become intimate, which created a divergence between the production of ideas and concepts of strategy and the implementation of strategy. In the world of corporate strategy, the existence of small and medium-sized firms that did not have the resources to run in-house strategy departments like Lockheed and the US government led to the development of the private strategy consultant. 'Perhaps the most important development in the history of strategy', writes Richard Koch, 'was the founding in 1963 of the Boston Consulting Group.' BCG started out with one man and one office, but in time Bruce Henderson was no longer alone. Koch estimates that from 1965 to 1991, the number of strategic consultants employed by the leading strategy consulting firms grew 15–20 per cent annually.[21] The strategic consultancies created a market for ideas and strategies that engaged academics, business people and consultants in processes of innovation. The end result is that the body of literature and scope of practice within corporate strategy is much larger and more end-user focused than the academic study of military strategy.

The difference between how corporate and military strategies have been institutionalised means that, in spite of the shared relevance and common heritage, the students of business strategy and military strategy neither write books for one another nor seem to read each other's books. Business scholars seem to be more aware of the existence of the other field of strategy, however, even if the occasional references to Sun Tzu and Clausewitz might seem a little clichéd to a scholar of strategic studies or military strategy.[22] Fortunately, this situation might be about to change. One reason is that the Western armed forces are increasingly using consultants to optimise their business practices, meaning that corporate strategy concepts more explicitly enter the military sphere; and since consultants significantly drive knowledge production within the corporation strategy field, the engagement with the military drives academic studies.

[21] Koch, *Strategy*, 8.
[22] Freedman, *Strategy*, 505–12.

In business studies, Stephen Bungay's work offers an excellent example of how military history and management can inform one another. Bungay explains the British victory in North Africa during the Second World War in terms of supply-chain management and advises business leaders to make a model of Moltke the Elder.[23] In military studies, Lawrence Freedman's *Strategy: A History* is the first textbook in the field to include business and military strategy side by side.[24] Still the application of business insights has, thus far, been patchy at best. Taken together, however, a number of very different studies constitute an agenda asking the questions of the armed forces which Constantinos Markides reasons should be asked of a business. Eliot Cohen has reintroduced the study of leadership in warfare in his *Supreme Command*.[25] Jacob Shapiro takes the management perspective further in his study of the organisation of terrorist groups arguing that 'the core managerial challenges of terrorist organisations are actually quite similar to those faced by other, more traditional human organisations'.[26] Steven Biddle describes how combined arms make up 'the modern system of employment' in order to be able to calculate the odds of success in military engagement.[27] Michael O'Hanlon builds on this ambition of measuring performance by reproducing the building blocks of staff work in terms of, for example, how many planes to allocate to a mission or the total cost of a heavy army division.[28] This is a very important fundament for a discussion about the business of the armed forces. Drawing on sociological literature on organisation and networking in a global, interconnected world,

---

[23] Stephen Bungay, *Alamein* (London: Aurum Press, 2002); Stephen Bungay, *The Art of Action: How Leaders Close the Gaps Between Plans, Actions and Results* (London: Nicholas Brealey Publishing, 2011).

[24] Freedman, *Strategy*.

[25] Eliot A. Cohen, *Supreme Command: Soldiers, Statesmen, and Leadership in Wartime* (New York: The Free Press, 2003).

[26] Jacob N. Shapiro, *The Terrorist's Dilemma. Managing Violent Covert Organizations* (Princeton University Press, 2013), 2.

[27] Stephen Biddle, *Military Power: Explaining Victory and Defeat in Modern Battle* (Princeton University Press, 2004).

[28] Michael O'Hanlon, *The Science of War: Defense Budgeting, Military Technology, Logistics, and Combat Outcomes* (Princeton University Press, 2009).

Anthony King approaches the European armed forces as a field of practice, or perhaps an industrial sector, and is thus able to draw a picture of the European NATO countries in a much more dynamic manner than the traditional list of military capabilities in the International Institute for Strategic Studies annual ledger of military capabilities.[29] Terry Terriff, Theo Farrell and Frans Osinga have launched a research project on innovation, which Osinga and Farrell have continued in studies of 'adaptation'. This defines a research agenda focused on how militaries adapt to operations that is similar to and informed by insights into how firms adapt to changing market conditions.[30] The focus on a transformation in the demand for military power is also prevalent in the writings on hybrid warfare, even if that literature is generally more interested in the reorganisation of the West's enemies than the reorganisation of Western armed forces to meet these new challenges.[31] Michael Alexander and Timothy Garden argue for a reorganisation of European armed forces based on pooling and sharing because 'the arithmetic of defence policy' dictates that this is the only way Europe can maintain a relevant military capability.[32] Research into defence economics, civilian–military relations, gender issues and other issues related to the organisation of armed forces are obviously relevant for these discussions but often treat the organisation of the armed forces themselves as a given, instead focusing on mapping the relationship between the armed forces and, for example, the defence industry.

This book utilises the insights of these studies to offer a comprehensive analysis of the business model for Western armed forces. The

[29] Anthony King, *The Transformation of Europe's Armed Forces: From the Rhine to Afghanistan* (Cambridge University Press, 2011).

[30] Terry Terriff, Theo Farrell and Frans Osinga, *A Transformation Gap: American Innovations and European Military Change* (Stanford University Press, 2010); Theo Farrell, Frans Osinga and James A. Russell (eds.) *Fighting the Afghanistan War* (Stanford University Press, 2013).

[31] Frank Hoffman, 'Hybrid Warfare and Challenges', *Joint Forces Quarterly* 52 (2009), 34–9.

[32] Michael Alexander and Timothy Garden, 'The Arithmetic of Defence Policy', *International Affairs* 77 (2001), 509–29.

first two chapters present an analysis of the current business model for armed forces. As such, readers familiar with military strategy will be surprised to find that the main focus is not on the opponents the armed forces are to gear up against but instead on how the armed forces are organised. Readers familiar with business studies will happily note that the analysis begins with an examination of the trends in, primarily, the US defence budget since 1948. This budgetary analysis is put in the context of the development in other Western armed forces, primarily the British, in order to demonstrate that the conclusions drawn from the American defence budget are – with some modifications – applicable to Western armed forces in general. When writing a book which at least in temperament had something in common with the present one, General Maxwell Taylor apologised in advance to the reader for discussing the intricacies of the US defence budget. 'This may appear to be a drab subject', Taylor wrote, 'but, if it performs its proper function, the budget is far more than a compilation of dusty figures of interest only to fiscal experts. It should be a translation into dollars of the military strategy upon which our future security will depend.'[33] I fear that comparing the US defence budget with the budgets of other nations as well as analysing tooth-to-tail ratios does not make the subject less drab. However, it renders Taylor's observation that the defence budget is 'the translation into dollars of the military strategy' no less poignant.

The predicament of the Western armed forces becomes apparent in the analysis of the budget numbers and the numbers in the tooth-to-tail ratio.[34] The modern military system may provide the framework for very efficient combined arms operations, and no one does that better than the US military. Measured in the system's own terms, however, this efficiency does not necessarily correspond to the efficient use of taxpayers' money. On the contrary, the armed forces

---

[33] Maxwell D. Taylor, *The Uncertain Trumpet* (London: Atlantic Books, 1959), xi.

[34] John McGrath defines the tooth-to-tail ratio as 'the number of troops in a military organization employed in combat duties versus the number functioning in noncombat roles'. John J. McGrath, *The Other End of the Spear: The Tooth-to-Tail Ratio (T3R) in Modern Military Operations* (Fort Leavenworth, KS: Combat Studies Institute Press, 2007), 2.

are increasingly ruled by the law of diminishing returns. Thoughtful military officers know this. They seek alternative business models that will make the military system more effective in targeting the nation's enemies while using the nation's resources more efficiently. Because of the size of the US defence budget and the US global commitments, this debate is driven by the United States. This does not mean that is not applicable to the allies. On the contrary, most European nations have their version of the American debates and take part in these debates in multinational research and development institutions such as the Allied Command Transformation, NATO Defense College in Rome or various think tanks across the Atlantic community. This debate on alternative business models assumes many different forms, but in general terms one can say that the debate is structured around two alternative business models.

Chapter 4 describes how a group of officers and academics – which I, with all due respect, refer to as 'the coinistas' – identifies demographic change and globalisation as the defining features in the twenty-first-century security environment. From this perspective, the wars in Iraq and Afghanistan came to define the shape of things to come and the counterinsurgency (COIN) strategy thus became not only a recipe for turning the tide in Iraq and Afghanistan but also a catalyst for change. The COIN approach is a human resource approach to organisational change, as it focuses on the individual soldier, airman or marine as the object of change. To the other business model for making ends meet in the armed forces, technology is the object of change (this is the subject of Chapter 5). From this perspective, the convergence of nanotechnology, biotechnology, information technology and cognitive technologies (NBIC) offers an opportunity to dramatically increase the efficiency and reduce the costs of military platforms. They are literally describing a *deus ex machina* that will allow the armed forces to remain platform-centric but with a far greater return on investment. Where the COIN approach found its concrete formulation in the Counterinsurgency Field Manual and the strategy for the surge in Iraq, the NBIC approach finds its preliminary, concrete application in the AirSea Battle doctrine.

However, neither the COIN approach nor the technological approach offers a way out of the logic of the modern military system. The result of applying these approaches as recipes for changing the military business model has been to increase the US defence budget. Even the most reformist officers thus seem unable – this is obviously written in very general terms – to be able to reframe and reprioritise the military system. At best, they appear able to add an element to the strategy and another item to the budget.

In order to actually reform the modern military system, I argue in the final chapter, the military's approach to strategy must be reconfigured. Instead of thinking of strategy in terms of means and ends, reform should be based on thinking of strategy as a design. The focus on strategic design informs this view, thus demonstrating the fruitful interaction between business strategy and military strategy. This approach recognises the insights in the COIN and technological approaches alike. Both point to important trends and both offer insightful responses to the consequences for the armed forces. However, both approaches fail to appreciate the extent to which technological change and global interconnectedness fundamentally reconfigure how strategy is done. Military strategy can no longer be regarded as a process carried out within the confines of the military system. This is possibly the most important reason for drawing on insight from business strategy, for the armed forces increasingly regard themselves as part of the government's entire security capacity, and the contribution of the armed forces to that capacity will be designed from operation to operation. No grand design will create the military system for the twenty-first century, but the armed forces will be in a continuous process of design.

2 The modern military system

'Pray explain, CIGS', Winston Churchill asked the Chief of Imperial General Staff, General Sir Alan Brooke, 'how is it that in the Middle East 750,000 men always turn up for their pay and rations, but when it comes to fighting only 100,000 turn up?'[1] Britain had recoiled from the retreat from Dunkirk and prevented a German invasion in the 'Battle of Britain'. In 1942, the Prime Minister's immediate fear was no longer that Britain would be defeated, but rather that the British armed forces were incapable of winning. In Singapore, a British army of 100,000 troops had been defeated by Japanese soldiers on bicycles, and the German *Afrika Korps* was pressing a superior British force to the brink in Egypt. What if the cold-blooded fact was that British soldiers and their commanding officers no longer knew how to produce a decisive victory on the battlefield? Even if Brooke was exasperated by the fact that much of his job was to take the great man's abuse, the General was no more impressed with the English fighting man than the PM. 'If the Army cannot fight better than it is doing at the present', Brooke wrote in his diary, 'we shall deserve to lose our Empire.'[2] Yet the answer to Churchill's pointed question about why most of the soldiers seemed to be in the rear rather than at the front was more complicated, and with consequences further into the future than either Churchill or Brooke could possibly appreciate at the time.

It is easy to dismiss Churchill's critique of the British Army's performance as yet another example of how his grasp of the past

---

[1] Churchill quoted in Stephen Bungay, *Alamein* (London: Aurum Press, 2002), 116–17. See also, Max Hastings, *Finest Years: Churchill as Warlord* (London: HarperPress, 2009), 259–72.

[2] Quoted in Hastings, *Finest Years*, 266.

was firmer than his grasp of the future. How was he to appreciate the nature of modern warfare? In military jargon, distinction is drawn between the 'teeth' of the army, which are the units on the front line in actual combat, and the tail, which takes care of administration, logistics and the like. Obviously, distinguishing between the two can be difficult, but it is a useful distinction when assessing the cost-effectiveness of the armed forces. In the Egyptian desert, the British deployed their most modern armoured formations. The 200–300 tanks of a British armoured division required a logistical tail much larger than ever experienced in the past, therefore the deployment in the desert only expanded the length of the tail.[3] This was not only true of the British. When the Allies were finally able to engage the Germans in Europe, only 39 per cent of the 2.1 million US troops in the European theatre of war were employed in combat units.[4]

Might Churchill's sense of history actually have allowed him to comprehend a problem, which the officer class, increasingly enmeshed in the mechanics of modern warfare, were unable to see as clearly? What if the 'tooth-to-tail' ratio actually demonstrated the bureaucratisation of modern warfare? To Churchill's immense relief, General Montgomery was able to turn the hapless army of the desert into the desert rats. Montgomery was only able to do so, however, by adopting a way of war that relied heavily on airpower, artillery and amour, so as to minimise the risks run by his own soldiers. Montgomery's success rested largely on the fact that he devised a way of winning which was suited to the reluctant British warriors.[5] If the British were reluctant warriors, they were at least enthusiastic

[3] Bungay, *Alamein*, 41–66.

[4] John J. McGrath, *The Other End of the Spear: The Tooth-to-Tail Ratio (T3R) in Modern Military Operations* (Fort Leavenworth, KS: Combat Studies Institute Press, 2007), 103.

[5] Bungay, *Alamein*, 207–10; Max Hastings, *Armageddon: The Battle for Germany 1944–45* (London: Pan Books, 2004), 588; Martin van Creveld, *Fighting Power: German and US Army Performance 1939–1945* (Westport, CT: Greenwood Press, 1974).

about military technology, and Churchill was keenly interested in military innovation.[6]

This capital-intensive means of warfare suited the industrial societies of Britain and the United States but came with a price tag – a price that only grew after the Second World War. Carl Conetta has thus calculated that the price per soldier per year was USD 393,000 in the Korean War (1950–53) and USD 256,000 per soldier per year in the Vietnam War (1961–73), whereas the price had risen to USD 792,000 in the wars in Iraq and Afghanistan (2001–15).[7] These numbers demonstrate that military budgets are to a large extent defined by the nature of particular conflicts. The Vietnam War was an infantry conflict par excellence with up to 500,000 American troops deployed over a long period, which obviously reduces the average cost, whereas the cost of the high-tech professional armies deployed in Iraq and Afghanistan was much higher. Despite these particularities, the increase in the cost of military force is a trend as powerful as it is obvious.

This chapter seeks to answer Churchill's question by analysing the armed forces in much the same way as one would analyse any other industry. Obviously, the US military is no ordinary company and an army brigade no factory. In spite of the unique nature of 'the military product' and the circumstances under which it is 'produced', any enterprise – public or private – must be able to justify its achievements in relation to costs. In the case of a public function so much at the heart of what a government does as the military, this question of efficiency cannot be answered in terms of quarterly returns, as would be the case in a private company. Instead, a long-term view is necessary in order to establish trends that are not just the reflection of certain geopolitical circumstances or the policy of a specific government. Taking the long-term view, one is also able get around the

[6] David Edgerton, *Britain's War Machine: Weapons, Resources and Experts in the Second World War* (Oxford University Press, 2011), 86–122.

[7] Carl Conetta, *An Undisciplined Defense: Understanding the $2 Trillion Surge in US Defense Spending*, Project on Defense Alternatives, PDA Briefing Report #20, 18 January 2010, www.comw.org/pda/fulltext/1001PDABR20.pdf (2 January 2013), 7.

fact that the giant sums involved in procuring an aircraft carrier or sustaining an army division in the field can distort the balances and tendencies within the budget when considering the defence budget in the short term. The long-term perspective also contributes to an understanding of the armed forces as an organisational type rather than focusing on the specific performance of one national military. In the context of this analysis, I am thus concerned with the performance of the organisation rather than victories or defeats on the battlefield. That is not to say that these are not important and in many ways the ultimate test of a nation's armed forces. Winston Churchill might have been justified in his fear that the Germans were the better soldiers, but Britain won the war anyway. The long-term perspective reduces the friction of particular wars and makes it possible to conclude something concerning the general viability of the military business model.

This chapter opens with a discussion of the paradox of military power – on the one hand, military power can deliver decisive results; on the other, these results do not always translate into the desired political results. The Germans came to learn this in the Second World War. This discrepancy between single-case success and overall success is important when focusing on what the military organisation achieves rather than on what a certain formation of armed forces achieves at a given point in time. Combat makes the military experience profoundly different from a civilian experience. Yet it also often stands in the way of evaluating the military business model. The first section of the chapter argues that such an evaluation is possible and very much needed. Having established that the armed forces can be dealt with as a business, the second part of the chapter proceeds to define a model of the military business. This is done by beginning with Stanislav Andreski's notion of a military organisation. Andreski had a Weberian approach to military organisation, arguing that ideal typical forms of military organisation could be defined. On this basis, I argue for the existence of a modern military system and use Stephen

Biddle's work to show how that system came into being and how it works.

THE MILITARY BUSINESS

The armed forces are not and cannot be a business; yet one can learn a lot about the armed forces by studying them in business terms. The all-important first step is to distinguish between regarding the armed forces as a business and using the lessons from business studies to get insights into how the armed forces work. To Adam Smith, maintaining armed forces was first on any government's to-do list. 'The first duty of the sovereign, that of protecting society from the violence and invasion of other independent societies, can be performed only by means of military force', Smith argued.[8] Military forces were thus first on Smith's list of government functions that the wealth of the nation had to support, followed by justice, public works and education. Smith was keenly aware, however, that even if the protection from violence and invasion was the government's business, this was a business that different governments undertook in different ways in the course of history. Thus, he went on: 'But the expense both of preparing this military force in times of peace, and of employing it in time of war, is very different in the different states of society, in the different periods of improvement.'[9] Even if the military is on any government's to-do list, governments do the military very differently. A study of history shows that the organisation and deployment of military forces reflects the nature of the society that engages them. As Clausewitz noted, 'every age had its own kind of war'.[10] A study of different militaries also demonstrates considerable differences in how they operate. These differences also include the extent to which military force is applied by entities that

[8] Adam Smith, *An Inquiry into the Nature and Causes of the Wealth of Nations*, (Indianapolis: Liberty Classics, 1981), 689.

[9] *Ibid.*

[10] Carl von Clausewitz, *On War*, edited and translated by Michael Howard and Peter Paret, (Princeton University Press, 1976 [1832–34]), 593.

are in fact private companies. The condottieri of Northern Italy in the Renaissance or the private military companies utilised by the US military today are private companies waging war for profit. Various support functions often rely on private contractors as well. Private companies may thus subcontract warfare from the state, but history also demonstrates that private companies themselves may field armies and navies, as did the British East India Company. The fact that these actors have a profit motive demonstrates that one has to distinguish between 'the first duty of government' and how this duty is carried out.

Military strategy often disregards this distinction. As mentioned, military strategy is focused on how to apply military power in certain circumstances. It focuses on how to deal with the situation at hand, and therefore military strategy privileges the application of capabilities over their production. Business strategy has the opposite focus. Business leaders have much wider scope for changing their organisation, and business strategy therefore deals with how to organise the business from production to sales. This means that military strategy and business strategy conceive of their agents in very different terms. The military strategist is focused on how the general applies his forces in the stress of battle, whereas the business strategist focuses on how the manager sets up an organisation and a production line that creates incentives for employees in a productive fashion. When business leaders turn Secretaries of Defence this creates a clash of cultures. Thus, Milan Vego of the US Naval War College cites the attempts to streamline the Department of Defense business model by Secretaries with business experience of their own, such as Robert McNamara (1961–68) and Donald Rumsfeld (1975–77, 2001–6), as proof that one cannot approach the armed forces as a 'business'. Vego believes that the reason why McNamara mismanaged the Vietnam War and Rumsfeld mismanaged the Iraq War was that these former business executives failed to perceive that 'the nature of war as explained by Carl von Clausewitz is not subject to change regardless of the changes in military technology, not to say the world's economy'.[11] Vego's statement demonstrates how the business focus on doing strategy by changing the ways of going about things military is in opposition to how the military regard the nature of their profession. Military men often regard war as a universal test which individual commanders must take in order to prove their worth. In Vego's argument, the fact that the military cannot be run like a business becomes proof that the military's business is forever the same. In this argument, the notion that the armed forces can be run like a business and the proposition that lessons from business can be used in running the armed forces become conflated into one. From this perspective, the organisational reality of the military and operational tasks are the same thing. In effect, this is a defence against the business approach to strategy; the ways of doing business cannot be changed, the argument goes, because the nature of the business is unchangeable.

Vego's argument reflects the unique mixture of machismo and denial that often defines the relationship between military professionals and the civilian world. Niklas Luhmann explains that to an organisation, the 'outside' only exists as a negation of the 'inside'. 'The unity of the environment is constituted by the system', Luhmann argues. 'The environment is only a negative correlate of the system. It is not a unit capable of operations; it cannot perceive, have dealings with, or influence the system.'[12] When considering the military as a system, then Luhmann can contribute to an understanding of how the difference between 'military' and 'civilian' is not really about the nature of civilian society, but rather about how civilian society is perceived by the military. In many ways, Vego's argument against business is an argument against an abstract 'Other' that will impose controls and efficiency measures on a military organisation 'it' does not understand. One indication of this is the absence of civilian society from the narrative. Civilian society is not an organisation, as is

---

[11] Milan Vego, 'Is the Conduct of War a Business?', *Joint Forces Quarterly*, 59 (4th Quarter 2010), 58.
[12] Niklas Luhmann, *Social Systems* (Stanford University Press, 1995), 182.

the military, but is nevertheless presented as a coherent entity. In that sense, civilian society becomes a golem constructed with all of the things the military is not or is not supposed to be; an assortment of characteristics which can subsequently serve as a mirror in which the military can see itself. Perhaps by necessity, that mirror is of the kind one would find in an amusement park, which distorts the actual reflection. A Secretary of Defense with a business background might hold up a mirror in which the armed forces seem fat compared to the lean businesses in the civilian sector. A career officer, on the other hand, may see an effective and dedicated workforce in the military, operating in a manner that could teach civilian society a thing or two about commitment. According to a 2003 survey, most of the US officers deployed to Iraq believed that civilian society would run better if run by military standards (more on this in Chapter 6).[13]

Vego argues that the military organisation is fundamentally different from civilian organisations, because the nature of the battlefield is irrational, whereas the nature of the marketplace is rational.[14] Carl von Clausewitz would agree that battle is 'the collision of two living forces'.[15] Clausewitz appreciated how the chaos of the battlefield was the result of giant organisations of thousands of men colliding under extreme conditions. When the cannonballs were flying and the bayonets fixed, the real challenge facing the commander was to avoid getting lost in the confusion and fury of the battlefield and keeping his objective in mind. Clausewitz thus distinguished between the irrationality of the battlefield itself and the rationality of warfare. 'War is a clash between major interests, which is resolved by bloodshed – that is the only way it differs from other conflicts',[16] he argued. As a veteran of the Napoleonic wars, Clausewitz had few illusions about the nature of the battlefield. For instance, he did not believe that fancy manoeuvring could allow the commander to

---

[13] Morten G. Ender, *American Soldiers in Iraq: McSoldiers or Innovative Professionals?* (New York: Routledge, 2009), 50.

[14] *Ibid.*, 59.   [15] Clausewitz, *On War*, 77.

[16] *Ibid.*, 149.

avoid engaging the enemy. 'Essentially war is fighting, for fighting is the only effective principle in the manifold activities generally designated as war', Clausewitz argued. 'Fighting, in turn, is a trial of moral and physical forces through the medium of the latter.'[17] The purpose of the armed forces was to provide units that could fight, Clausewitz concluded. 'The whole military activity must therefore relate directly or indirectly to the engagement. The end for which a soldier is recruited, clothed, armed and trained, the whole object of his sleeping, eating, drinking, and marching is simply that he should fight at the right place at the right time.'[18]

Vego states that another reason why the armed forces cannot be governed by a business logic or led by Secretaries of Defense fancying themselves as business executives is that soldiers live closer together and form a 'band of brothers' that is foreign to civilian life.[19] It seems undeniable that a platoon in combat constitutes a very special community of fate. But does it necessarily follow that the entire military workforce has the same relationship as the combat platoon and that the entire military organisation works according to the same 'band of brothers' logic as the combat platoon? Is the lieutenant colonel working with human resources in the Department of Defense really doing something fundamentally different from what an HR administrator in a large civilian corporation would be doing? Clausewitz would probably not agree. Clausewitz did not believe the entire armed forces to be governed by the same logic as front-line units. 'An interaction exists between the preparations and equipment and fighting', he noted, 'but fighting itself still remains a distinct activity; the more so as it operates in a peculiar element – that of danger'.[20] The medium of warfare was fighting, according to Clausewitz; but he did not believe that combat was the purpose of warfare. War was to deliver political results, and the armed forces were organised in order to best deliver those results. In that sense, the military provided a service for the

---

[17] *Ibid.*, 127.   [18] *Ibid.*, 95.
[19] Vego, 'Is the Conduct of War a Business?', 59.
[20] Clausewitz, *On War*, 127.

sovereign. Actually, Clausewitz was fond of comparing the military to a business. He compared war to 'commerce', arguing that combat is the medium of war in the same way as money is the medium of commerce.[21] *On War* is perhaps the first book that tries to describe and define war with a civilian reader in mind, and Clausewitz clearly found that the business analogy offered a way of conveying to the German *Bürger* reading his book that while the bloody events of a war were very different from everyday Berlin life, the logic of the battlefield was little different from that of the marketplace. War was fundamentally an interaction between two sides that valued their results not in terms of profit but in terms of political gains.

Clausewitz's insight that an army is much more than a unit of battle was continued in Field Marshal Helmut von Moltke's *Guidance for Large Unit Commanders*. If *On War* provides a theoretical argument for why the business of war cannot be reduced to battle, then Moltke's army reforms demonstrate that even if modern military strategy often presents a reified picture of the nature of warfare and the utility of armed forces, this has not always been the case. Moltke issued his *Guidance* for the Prussian army in 1869. The Prussian army had defeated Austria in 1866 at the battle of Königgrätz, establishing Prussia as the predominant German power. If this predominance was to be used to achieve the long-term strategic aim of uniting Germany under Prussian leadership, the army would have to fight again; and Moltke was not a man prone to rest on the laurels of victory. Stephen Bungay points out how the Field Marshal was in a situation similar to that of thousands of executives in large corporations today.[22] The organisation had a clear vision for what to do; the real challenge was how to do it. Moltke was able to conceive of the army as an organisational entity at the same time as he obviously regarded it as a means to fight wars. 'The field of real activity for an army is war', Moltke argued. 'Nevertheless, its development, its

---

[21] *Ibid.*, 149.

[22] Stephen Bungay, *The Art of Action: How Leaders Close the Gaps Between Plans, Actions and Results* (London: Nicholas Brealey Publishing, 2011), 54–110.

normal state, and the largest portion of its life fall in times of peace. This contradiction renders goal-oriented training difficult and raises the danger of a sudden transition [to war]'.[23] Moltke thus focused on developing the capacity to deal with sudden transition in the army as such as well as in the individual officer. The nature of military conflict was such that a detailed plan was doomed to fail. Information about the battlefield was often not available beforehand, and even then, the 'friction' of the fight would soon render that information obsolete. The battle developed more quickly than it was possible to transmit information from the front to the commander and instructions back to the front-line units.

Moltke's solution was to place the responsibility for acting on to the front-line units that actually had the information. This was the first point of his *Guidance*. The second point was to ensure that the individual front-line units would use their superior tactical information to achieve the overall strategic objective. The Prussian army called this *Austragstaktik*. For Moltke, strategy was about leadership. This focus on action largely stemmed from the notion of the officer as a romantic hero who has an intuitive notion of when to strike, which can be found in Clausewitz's *On War*. 'In the conduct of war what one does often matters less than how one does it', Moltke argued.[24] Moltke's guidance focused officers' attention on the 'centre of gravity' and placed trust in their ability to execute an offensive on that centre of gravity. Thus, it is only a slight overstatement to claim that it was human resources that won the Franco–Prussian War (1870–71) for Prussia, rendering it possible for the Field Marshal to realise Bismarck's strategic objective of German reunification. The focus on the autonomy of field commanders, however, is another reason why Germany lost the First and Second World Wars. *Austragstaktik* created an organisational culture focused on tactics

---

[23] Helmuth von Moltke, '1869 Instructions for Large Unit Commanders', in Daniel J. Hughes (ed.), *Moltke on the Art of War: Selected Writings* (Novato, CA: Presido Press, 1993), 172.

[24] *Ibid.*

and operations rather than on strategy. With the political leadership unable to define clear strategic aims, the German army focused on immediate tactical results rather than long-term strategic gains.[25]

Moltke's example shows how great military leaders are able to combine the organisational focus of business strategy and the focus on prevailing in particular circumstances, which is the focus of military strategy. Moltke had a keen awareness that 'a great deal of strategy work is trying to figure out what is going on', which is Richard Rumlet's description of what he finds to be the most important part of strategising.[26] The kernel of strategy, according to Rumlet, is to diagnose the situation; a diagnosis which includes one's own capabilities as well as those of one's opponent. Following this line of reasoning, Lawrence Freedman argues that 'strategy starts with an existing state of affairs and only gains meaning by an awareness of how, for better or worse, it could be different'.[27] This appreciation that things might be different is simply not possible if one reifies the nature of the military and military power. In that case, strategy becomes a question of application rather than reflection. This was exactly what Moltke feared the most; an officer corps that was only able to apply plans but not to reflect on what to do if the plans did not match up with the realities of the battlefield would simply not be conceptually equipped to prevail – no matter how well supplied with guns and logistics they might otherwise be. In this respect, at least, Moltke thought more like a modern businessman than a modern general.

The reification of the military not only has consequences for how one regards the military organisation but also for how one imagines that organisation can be used. 'It is tempting if the only tool you have is a hammer to treat everything as if it were a nail',[28] Abraham

[25] Holger H. Herwig, *The Marne – The Opening of World War I and the Battle That Changed the World* (New York: Presidio Press, 2012).

[26] Richard Rumlet, *Good Strategy/Bad Strategy: The Difference and Why It Matters* (London: Profile Books, 2011), 79.

[27] Lawrence Freedman, *Strategy: A History* (Oxford University Press, 2013), 611.

[28] Abraham H. Maslow *The Psychology of Science* (New York: Joanna Cutler Books, 1966), 15.

Maslow famously noted. This is also true of the armed forces. Steven Walt notes that 'when you've got hundreds of planes, smart bombs, and cruise missiles, the whole world looks like a target set'.[29] When the nature of military power is taken for granted, then how a government can deal with national security issues is taken for granted as well. Armed force is applied in accordance with a template defined by the military itself rather than the problem at hand. This lack of inter-relationship between problem and solution stems from the fact that military force is often not regarded as a solution to a given problem but rather as an action in and of itself. When the armed forces are not regarded as an integrated part of a national strategy that also includes civilian measures, then military operations would have no more point than a training exercise if it were not for the cost in human life. Lecturing a military audience at the US Naval War College in Rhode Island, General Tommy Franks was asked about the nature of the war in Afghanistan, which Franks directed as the commander of the US Central Command. 'That is a great question for historians', General Franks answered.[30] Franks simply did not have the feel for strategy as the understanding that something could be different. To the General, battle was something that had to be joined rather than understood and shaped in order to fit the overall national aim. It was a campaign rather than a continuation of policy, and Franks therefore only had tactical answers to the quintessential strategic question.[31] Adopting such a tactical approach makes it equally difficult to assess how new technologies and operational concepts can be used to reconfigure the conduct of operations.

Such questions concerning innovation would be central in business strategy, but in the domain of military strategy they are seen as

<hr>

[29] Stephen M. Walt, 'Is America Addicted to War?', *Foreign Policy*, 4 April 2011, www.foreignpolicy.com/articles/2011/04/04/is_america_addicted_to_war?page=full (13 November 2012).

[30] Thomas E. Ricks, *The Generals: American Military Command from World War II to Today* (New York: Penguin, 2012), 400.

[31] Hew Strachan, *The Direction of War: Contemporary Strategy in Historical Perspective* (Cambridge University Press, 2013), 68–70.

factors to enable operations rather than elements that can reconfigure how operations are done. The result is conceptual inertia. 'Virtually all present-day discussions of conventional warfare are still conducted in terms originally coined by the likes of Guilio Douhet, Eric Ludendorff, Hans von Seeckt, John Fuller, and Basil Liddell Hart in the years before 1939',[32] Martin van Creveld notes in what is only a slight overstatement. Military strategy is discussed not by a logic of consequentiality, but in terms of a logic of appropriate action according to which action 'involves fulfilling the obligations of a role in a situation, and so of trying to determine the imperatives of holding a position'.[33]

This logic of appropriateness was at work when President Obama went to the Pentagon on 5 January 2012 to announce large defence cuts. In the Pentagon briefing room, the President told the press that 'the United States of America is the greatest force for freedom and security that the world has ever known. And in no small measure, that's because we've built the best-trained, best-led, best-equipped military in history – as Commander-in-Chief, I'm going to keep it that way.'[34] After the speech, Deputy Secretary of Defense Ashton Carter explained that the Defense Department would be cutting USD 489 billion over the 2012–22 period, with additional savings in the 2012–17 period of USD 261 billion.[35] In the following years, defence cuts continued, with the Defense Department describing this as 'a period of increasing fiscal constraint'.[36] The President's remarks were designed to make less of these cuts by arguing for them in terms of appropriateness. The cuts were appropriate from a budget

[32] Martin van Creveld, *Technology and War: From 2000 BC to the Present* (New York: The Free Press, 1991), 277.

[33] James March and Johan P. Olsen, *Rediscovering Institutions: The Organizational Basis of Politics* (New York: The Free Press, 1989), 160–1.

[34] Barack Obama, *Remarks by the President on the Defense Strategic Review*, The White House. 2012, www.whitehouse.gov/the-press-office/2012/01/05/remarks-president-defense-strategic-review (20 January 2012), §1.

[35] Daniel Wasserbly, 'Obama Looks to Leaner Military, As-Pac Focus', *Jane's Defence Weekly*, 49 (2012), 4.

[36] US Department of Defense, *Quadrennial Defence Review 2014*, March 2014, www.defense.gov/pubs/2014_Quadrennial_Defense_Review.pdf (28 March 2014), iv.

perspective, and Obama wanted to demonstrate that he conducted himself presidentially by cutting without undermining the military capability of the United States. The cuts were mere technocratic necessities left to a Deputy Secretary. The message the President wanted to send was that these cuts would not affect the quality of the US armed forces. The United States was the world's leading military power, President Obama assured the electorate, which would surely hear that he was weakening the US military from the opposition in Congress. Actually, the US military was not only the best in the present world, the President argued, but the best military force in history. As discussed below, being number one was an important part of the US grand strategy, and the President assured the public that he would keep it that way despite the cuts. In other words, it was not so much what the US armed forces could do as the appropriate way of dealing with them that was in focus.

In his speech at the Pentagon, President Obama went on to explain that the cuts in defence spending were driven by careful, strategic considerations, 'because the size and the structure of our military and defense budgets have to be driven by a strategy, not the other way around'.[37] The President not only promised that the United States would remain the world's dominant military power, he also assured the public that his team knew what force structure was needed to produce superior military power. The main message of the review justifying the cuts was a strategic focus on Asia–Pacific, meaning continued investment in the most capital-intensive parts of defence in order to implement the Navy and Air Force's notion of AirSea Battle (more on this in Chapter 5), while cutting the number of troops in the Army. Yet the document listed a long number of priorities which appeared to reflect an undiminished demand on the US armed forces.[38] In Obama's strategy, the need to be the number-one

---

[37] Obama, *Remarks by the President on the Defense Strategic Review*, §8.

[38] US Department of Defense, *Sustaining US Global Leadership: Priorities for 21st Century Defence*, January 2012, www.defense.gov/news/Defense_Strategic_ Guidance.pdf (20 January 2012).

military power, and thus to be able to address strategic challenges across the board, was clearly at odds with the interest in prioritising resources. Assessing the Obama administration's national security strategy, Rosa Brooks wryly notes that '"long" isn't the same as "grand"'. The 60-page national security strategy contains many aspirations but no grand, coherent vision.[39]

The logic of appropriateness also works in the way the media doomsday machine works overtime, describing the calamities resulting from budget cuts. This puts pressure on the President and his Secretary of Defense to find the savings in other parts of the federal budget. The general economic malaise made such political priorities difficult, however. In order to bring government spending under control, Congress and the President had agreed on automatic savings to be implemented in all departments unless Congress was able to agree on prioritised spending cuts. Faced with the prospect of further cuts to his budget, Secretary of Defense Panetta warned that the US military risked becoming a 'paper tiger'. 'We would have to formulate a new security strategy that accepted substantial risk of not meeting our defense needs', Panetta argued, 'a sequestration budget is not one that I could recommend.'[40] *The Economist* pointed out that the logic of the argument was 'always more, or else'.[41] The Department of Defense executives have convinced themselves that there is only one appropriate way to go about the military's business. This comes with a price tag, which of course might be adjusted to fit political desires and financial realities but which fundamentally buys a fixed measure of security at a fixed price. It illustrates how any organisation's perception of its own standing and prospects for future development

---

[39] Rosa Brooks, 'Obama Needs a Grand Strategy', *Foreign Policy*, 2012, www.foreignpolicy.com/articles/2012/01/23/obama_needs_a_grand_strategy (24 January 2012).

[40] FOX News, 'Panetta Warns of Smallest Air Force Ever if Deep Defense Cuts Made', 20 January 2012.

[41] *The Economist*, 'Defence Spending: Always More, or Else', 1 December 2011, www.economist.com/blogs/democracyinamerica/2011/12/defence-spending (20 January 2012).

are vulnerable to misperception when not judged in terms of consequence but in terms of appropriateness.

In 1976, at a time when priorities and defence budget funding were similarly challenged, Andrew Marshall noted that 'we may well be at a point where we need to closely examine the very nature of the "business" we are in'.[42] This way of thinking represents a break with the logic of appropriateness and a challenge to scrutinise what the armed forces actually achieve. Adopting this 'logic of achievement' is a different way to approach strategy than the one most often associated with military strategy. This approach is much more common within the study of business strategy. Military strategy usually focuses on the application of an organisation's (the armed forces) resources to deal with a certain problem, whereas business strategy usually focuses on how to frame an organisation to best deal with a certain problem. The best commanders and most insightful strategists have of course been able to combine the two – this is one of the reasons why we still read Clausewitz and Moltke. One must appreciate how the modern military is organised – the modern military system – in order to understand the fiscal and organisational realities behind the logic of appropriateness and move towards a logic of achievement. Studying the military's business is not about studying the armed forces as if they were a private company operating in a free market, but about adopting the logic of achievement in focus by business studies. It is to study the military not in terms of military strategy but in terms of business strategy. The next section analyses the background for the modern military system.

THE MODERN MILITARY SYSTEM

Stanislav Andrzejewski acquired his first books by Max Weber in trade for food and cigarettes from German civilians when he was part

---

[42] Andrew Marshall, *Strategy for Competing With the Soviets in the Military Sector of the Continuing Political-Military Competition*, Department of Defense Office of Net Assessment, 1976, mimeo, http://goodbadstrategy.com/wp-content/downloads/StrategyforCompetingwithUSSR.pdf (20 January 2012), 8.

of the British army of occupation in Germany following the Second World War. Andrzejewski (or Andreski as he became known in Britain) was born in 1919 to a Polish merchant family. In 1939, he was studying economics at Poznan University when he was mobilised as an officer cadet to be part of the Polish defence against the German and Soviet invasions. Andreski was quickly captured by Soviet troops but managed to escape and avoided the fate that befell the other Polish officers at Katyn. Together with a friend, Andreski crossed the border to Slovakia and continued to Hungary. He eventually arrived in London, where he joined the Polish forces in exile and resumed his studies in economics at the London School of Economics. At the LSE, he met Karl Mannheim, who encouraged the young scholar-soldier to study Max Weber; Andreski got a chance to acquaint himself with the work of the great German sociologist when his unit was posted in Germany as part of the British army of occupation, and this was where Andreski was able to trade food and cigarettes with German civilians for Weber's books.[43] 'During the war', he notes in the preface to *Military Organization and Society* with an understatement that is as English in character as it is Central European in experience, 'I had the opportunity to observe many different armies, sometimes from a rather unusual point of vantage, and to ponder over the problems of military organisation.'[44]

The salient point of Weber's sociology is that modern societies are defined by certain organisational types which provide an overarching logic and, thus, societal cohesion in an otherwise individualised society. Weber's contemporaries, such as Tönnies and Marx, believed that modern society was evolving from an organic, integrated past to a mechanical, alienated future. In the many different companies on the market and in the individual career paths independent of kinship and traditional loyalties, however, Weber saw structures that defined

---

[43] 'Stanislav Andreski: Forthright Founder of Reading University Sociology Department', *The Guardian*, 7 November 2007.

[44] Stanislaw Andrzejewski, *Military Organization and Society*, 2nd edition (London: Routledge & Kegan Paul, 1968), xi.

individual action rather than structures that crush individuality. Weber observed that the modern organisation was structured on the basis of a bureaucratic organisational principle and that this principle in turn was guided by a 'means–end rationality'.[45] This allowed Weber to think in terms of types. No matter how different a chemical factory, a department store and a government department might seem, they were in fact very similar in terms of their organisational mechanics, Weber pointed out; 'for substantive reasons we may hope to facilitate the presentation of an otherwise immensely multifarious subject by expediently constructed rational types'.[46] Reading Weber's sociology amongst the ruins of post-war Germany, Andreski must have been struck by the similarity between the British, American, French and Soviet armies of occupation. The huge differences in funding, organisation and operational experience between these vast organisations dissolved when they were regarded as concrete manifestations of a Weberian ideal type. This is the fundamental insight Andreski draws upon in *Military Organization and Society*, which he published in 1954 when he had begun the career as a sociologist that would eventually lead him to become the head of the Sociology Department at the University of Reading for twenty years.

Andreski's book was one of the first to regard the military organisation as a sociological subject and in many ways started the sub-discipline of military sociology. 'Success in war, more than any other human activity', Andreski noted, 'depends on co-ordination of individual actions, and the larger a group the more necessary is the co-ordination, and the larger the hierarchy required.'[47] Thus, in Andreski's view, the defining factor in the constitution of the military as an organisation is how large a part of the society's overall

---

[45] Stephen Kalberg, 'Max Weber's Types of Rationality: Cornerstones for the Analysis of Rationalization Processes in History', *The American Journal of Sociology* 85 (1980), 1145–79; Ann Swidler, 'The Concept of Rationality in the Work of Max Weber', *Sociological Inquiry* 43 (1973), 35–42.

[46] Max Weber, *From Max Weber: Essays in Sociology*, H. H. Gerth and C. Wright Mills (eds.) (London: Routledge, 1997), 324.

[47] Andrzejewski , *Military Organization and Society*, 29.

resources, mostly in terms of labour, it commands. He describes this as the military participation ratio (MPR), which he defines as 'the proportion of militarily utilized individuals in the total population'.[48] This is where Andreski's reading of Weber becomes most clear. The MPR enables him to identify different types of military organisation based on how different MPRs have produced different military organisations and different civilian–military relationships. These types are essentially 'business models'. Slack and Lewis define a business model as 'a high-level design of the organisation that defines the structure and a style which enables it to meet its business objectives'.[49]

The military business model can be used in the same way Weber used his ideal type of bureaucracy. The model describes how certain objectives are being pursued by organisations designed in a certain way. 'War in our time is a war of machines', Max Weber noted, 'and this makes magazines technically necessary, just as the dominance of the machine in industry promotes the concentration of the means of production and management.'[50] Magazines must be managed, which inevitably leads to the bureaucratisation of the military. The crucial point of modern warfare, however, is that machines are used in combination with infantry and other machines. Prevailing in combined warfare is the objective the modern military business model is to achieve.

Stephen Biddle describes how this is done in his 'modern system of employment'. According to Biddle, this system was the answer to the misfit between man and machine that became terribly clear during the First World War. A Napoleonic infantry battalion with flintlock muskets could fire 1,000 rounds a minute; a 1916 infantry battalion armed with magazine rifles and supported by four machine guns could fire 21,000 rounds per minute.[51] Before the war, students

---

[48] *Ibid.*, 33.

[49] Nigel Slack and Michael Lewis, *Operations Strategy*, 3rd edition (Harlow: Prentice Hall, 2011), 7.

[50] Weber, *From Max Weber*, 221.

[51] Stephen Biddle, *Military Power: Explaining Victory and Defeat in Modern Battle* (Princeton University Press, 2004), 29.

of industrial society such as Friedrich Engels and Jan Bloch had predicted that war would become industrialised slaughter – a point echoed in the science fiction of H. G. Wells.[52] When the actual fighting commenced, there was very little the generals could do but to act out this horrible script, because they had little experience in how to use the tremendous firepower at their disposal in a battle against a similarly equipped force.[53]

Modern artillery and machine guns were able to stop the German offensive in the summer of 1914, and the war settled into a long bloody slog with trenches stretching from the Swiss border to the Channel.[54] The trenches turned out to be one of the greatest experiments in modern warfare. The Allies tried to break the deadlock by shelling the German trenches until the opposition was so devastated that the trenches could be taken unopposed by the infantry – *l'artillerie conquiert, l'infanterie occupert*, as the French said at the time.[55] At the Battle of Messines in July 1917, the Allies dropped 1,200 tons of explosives on every mile of the German line, equating to more explosive in terms of bomb per mile than a tactical nuclear weapon.[56] The Germans responded by developing a defence in depth, which left the first line of trenches to be shelled while further lines remained to absorb the Allied attack that followed the initial bombardment. This flexible approach to troop deployment was experimented on further by the Allies and Germans alike. The side that could first develop a way of combining the industrial power of artillery with an offensive version of the German defences would win the war. In their last gamble for victory – Operation Michael – in 1918, the Germans combined artillery and infantry in an offensive that almost overran the Allied positions, but in the end it was the Allies who got the system right in what the British referred to as the

---

[52] Christopher Coker, *War and the 20th Century* (London: Brassey's, 1994), 98–125.

[53] John Keegan, *The First World War* (London: Vintage, 2000).

[54] John Keegan, *The Face of Battle: A Study of Agincourt, Waterloo and the Somme* (London: Pimlico, 1991).

[55] Biddle, *Military Power*, 32.

[56] *Ibid.*, 30. On the soldiers' experience of these, see Keegan, *Face of Battle*, 204–84.

'All Arms Battle' and pushed the Germans back 200km,[57] and the German high command realised that the war was over.[58] The end of the war was the beginning of a new military system, Biddle argues:

> A remarkably stable and essentially transnational body of doctrinal ideas emerged from the crucible of the First World War. These ideas turn on reducing exposure to the radical firepower of the modern battlefield and enabling friendly movement while the slowing the enemy's. Taken together, these methods broke the trench stalemate in 1918 and defined the standard for successful military operations throughout the post-1918 era.[59]

What the modern military system really does is to transmit information and resources in order to amass force at a decisive point, thus achieving what Biddle terms 'preponderance'.[60] Donald Vandergriff aptly describes this as 'synchronization warfare'.[61] To be able to wage synchronisation warfare, commanders required balanced forces capable of being combined into a decisive strike force at the right time at the right place. Since the mid twentieth century, an army has needed a mix of amour, infantry, artillery and air power to be successful, while a navy has needed various classes of surface ships, submarines and aircraft to prevail. A US aircraft carrier group offers an example of such a combined force at sea, while a US infantry division is an example of combined arms on land. On the strategic level, the modern system requires a mixture of what the Department of Defense's

---

[57] Biddle, *Military Power*, 34; Holger H. Herwig, 'The Battlefield Revolution, 1885–1914', in MacGregor Knox and Williamson Murray (eds.), *The Dynamics of Military Revolution 1300–2050* (Cambridge University Press, 2001), 114–31; Peter Hart, *1918: A Very British Victory* (London: Phoenix, 2009).

[58] Martin Kitchen, 'Ludendorff and Germany's Defeat', 51–66, in Hugh Cecil and Peter H. Liddle (eds.), *Facing Armageddon: The First World War Experienced* (London: Leo Cooper, 1996).

[59] Biddle, *Military Power*, 190.

[60] *Ibid.*, 69–72.

[61] Donald Vandergriff, *The Path to Victory: America's Army and the Revolution in Human Affairs* (Novato, CA: Presidio Press, 2002), 9.

budget refers to as strategic forces, general-purpose forces, mobility forces, National Guard, special forces and reserve forces.

In order to be able to amass a decisive force at a decisive moment, the modern system of force employment requires a massive organisation that enables the combat forces to engage. During the stalemate in the trenches, neither the German nor the Allied infantry divisions were able to deliver decisive outcomes because they were not placed in a position to do so. The modern system of force deployment seeks to create these conditions. The forces the British Army deployed to the Middle East in the 1940s were thus very different from the forces Churchill commanded in his brief spell as a colonel of the 6th Royal Scots Fusiliers at the front in France in 1916.[62] These forces relied heavily on logistics and other enablers in order to be able to engage the enemy at the right time and place. As Churchill complained to General Sir Allan Brooke, this had profound consequences on the tooth-to-tail ratio. McGrath notes that noncombat roles are not merely logistics but also include administration, public relations, people who runs camps, military bands, etc.[63] The modern system of force deployment makes these members of the military bureaucracy the crucial players. Clausewitz had opined that the real challenge of military strategy was for the commander to assert his will in the chaos of the battlefield. Ernest Jünger described the modern battlefield as a 'storm of steel',[64] and the commander needs ways of processing information about the battle in order to direct an army in that storm. When victory became a matter of timely and coordinated deployment, information in the form of command and control became even more important.

Niklas Luhmann's notion of systems is helpful to understand what the modern system of force deployment in fact does. Luhmann begins with the fact that the 'the environment is always more

[62] Roy Jenkins, *Churchill: A Biography* (London: Pan Books, 2002), 299–309.

[63] *Ibid.*, 5–8.

[64] Thomas Nevin, 'Ernest Jünger: German Stormtrooper Chronicler', in Hugh Cecil and Peter H. Liddle (eds.), *Facing Armageddon* (London: Leo Cooper, 1996, 269–77).

complex than the system itself'.[65] During the First World War, the complexity of the modern battlefield overwhelmed the commanders because they did not grasp the system of deployment that could overcome the 'storm of steel'. They were – to continue the metaphor – shipwrecked on the battlefield without knowing how to get to shore. The modern system was a way of getting the ship back in the water and weathering the storm by imposing order not on the battle but on the armies. The vain attempt to force a decisive outcome by infantry charges against the enemy's trenches was – from a Luhmannian perspective – a way of trying to control the battle itself. Clausewitz knew that was not possible and that the true test of military genius was the ability to act with discipline and purpose in the chaos of warfare. The modern system rested on the tacit acknowledgement that a modern army was unable to control its environment. As Luhmann notes, a system 'must compensate for its own inferior complexity by superior order'.[66] The disorder of the battlefield was mirrored in the order of the modern system of deployment. Thus, the bulk of the armed forces' resources were concentrated on creating order by developing a tail that could deliver troops, equipment and supplies at the right time and place. In the tail of the armed forces, the logic of bureaucracy could create the purpose and discipline that the 'storm of steel' denied on the battlefield. Thus, the modern system creates what Luhmann terms 'schemata', which define procedures for how to go about war.[67]

Luhmann's crucial point about systems is that they are created to solve a certain problem (like the modern system of military deployment was created to prevail in the First World War), but once created, systems become a reality in their own right. 'Complex systems must adapt not only to their environments but also to their complexity', Luhmann notes[68] and goes on to argue that 'complex systems are forced to adapt to themselves'.[69] Once the military system was

[65] Luhmann, *Social Systems*, 182.   [66] *Ibid.*, 182.
[67] *Ibid.*, 187.   [68] *Ibid.*, 31.   [69] *Ibid.*, 32.

established in its modern form, the key to victory was to prevent the enemy's system from working. When the German army returned to northern France in 1940, it was able to defeat the French army using its *Blitz Krieg* tactics, which paralysed the French army system even if the army had plenty of remaining combat-ready units. The French were simply unable to deploy their forces.[70] The complexity of the military system not only defines the relations between individual armies, however. This complexity also defines the internal operations of individual armed forces. Luhmann points out that systems must organise their own internal problems as well as deal with external ones. Thus, a system is not merely a matter of input, output and feedback; it has its own logic that defines its own internal struggles. This is essentially the point Vego makes – the military has its own way of doing things. The business model of the modern military system, however, defines how the military operates.

By studying the armed forces in terms of the modern system of force deployment, one can pursue Andreski's insight that, for all the chaos of the battlefield and all the cultural differences of particular military organisations, armed forces can be studied as Weberian ideal types. Biddle convincingly argues that modern warfare challenged military organisation to find a mode of organisation that made decisive outcomes possible even in the 'storm of steel' on the modern battlefield. The military system consists of four elements: (i) combined arms operations, (ii) capital-intensive forces, (iii) focus on 'the tail', that is the logistical and administrative basis of operations and (iv) the flow of information that allows command and control over the individual elements of the system and facilitates their timely deployment.

It is important to note that this system was developed in competition between the armed forces of the great powers. An innovation by one side was readily adopted by the other and tested in battle until the modern system appeared in mature form. The mature form

---

[70] Alistair Horne, *To Lose a Battle: France 1940* (London: Penguin, 2007).

of the military system requires enormous resources in order to field a balanced force of combined arms with a system to facilitate its deployment. Thus, while many armed forces operate with the modern military system, it is not realised at full scale anywhere but in the US armed forces. Only the American government has at its disposal the combined arms in three services able to carry out the full range of operations according to the military system. From this perspective, the US armed forces represent the realisation of the ideal type that Biddle describes in the modern system of force deployment. The modern military business model is realised to perfection in the American military, for which reason it is natural to begin with the US armed forces when studying the present and future of the military business model. One of the reasons why the United States has these capabilities is that – in absolute terms – the United States spends more on defence than the combined budgets of the nine largest military spenders (the United States excluded, of course).[71] In that sense, the US military dominates the 'global military market' and thus sets the standards for how the modern military system is implemented in practice.

The size of the US military means that the US armed forces have in many ways conducted a 'horizontal expansion' by taking over the other 'firms' in the military 'industry'. Through NATO and other arrangements for military cooperation, the US military has set the standards for how the forces among its allies should develop. In turn, this has defined the standards for the weapons industry as well as the standards by which public and private consultants have tried to get the armed forces of the Global South into shape. The aggregated result of this has been the proliferation and entrenchment of the modern military system. That is not to say that, for example, the Danish and Japanese militaries are the same, but it is to argue that they operate along the same business model and are not competing. Rather, they are to be regarded as franchises in a global military enterprise

---

[71] International Institute for Strategic Studies, *The Military Balance* (London: Routledge, 2012), 31.

much in the same way as McDonald's restaurants in Copenhagen or Tokyo are part of the same enterprise. Where the US military thus can be said to set global standards in the sense that friends and foes alike neglect the American way of going about military business at their peril, the US allies, especially those in NATO, constitute an even more integrated community in which the armed forces of the participating nations are even more similar. From this perspective, it makes sense to speak of Western armed forces as a single object of study, even if doing so requires great care with respect to national differences in the organisations and funding of the armed forces.

Thus far, this chapter has established that one can study the military business model and how it is defined by what Biddle calls the modern military system. Using the sociology of Andreski and Luhmann, the modern military system has been defined in terms of four elements: (i) combined arms operations, (ii) capital-intensive forces, (iii) focus on 'the tail', the logistical and administrative basis of operations, and (iv) the flow of information that allows command and control over the individual elements of the system and facilitates their timely deployment. Furthermore, it has been established that the US armed forces realise the modern military system most fully. In the following, the US military will be the point of departure for the discussion concerning the present state of the modern military system and how that system presents challenges for using military force.

3 The modern military system: the US case

The US military is heavy. It is heavy in the sense of military jargon in that it is organised according to combined arms doctrine, and it is heavy in the sense of owning a lot of stuff that weighs a lot. The US military is also heavy in the more popular sense of the word, however; it is a large, complex, awe-inspiring organisation. A few examples demonstrate how heavy the US military actually is on all three counts. A heavy US Army brigade needs approximately 600 tons of supplies per day in combat. The 3,100 soldiers in the brigade require 300 tons of fuel, 130 tons of water, 85 dry stores and 60 tons of ammunition.[1] An Air Force combat wing uses up to 2,000 tons of ground-support equipment, including 100–200 tons of ammunition per day and up to 3,500 tons of fuel a day.[2] Transporting these forces and their supplies represents a challenge unto itself. To supply an Army brigade with 600 tons of supplies per day roughly requires 100 trucks.[3] A major road in good condition can perhaps handle 2,000 truckloads a day.[4] Transporting the US armed forces to other places on the globe is thus a major issue in determining 'the how' of the military business model. Even the US armed forces cannot easily be deployed beyond a well-functioning infrastructure. Good roads, size-able harbours and large airfields are a precondition for deployment. However, even a large airfield cannot manage more than 1,000 tons of supplies per day.[5] One way to work around this issue is to maintain an overseas presence. If the US Navy was operating from American ports instead of overseas bases, some estimates conclude that it

---

[1] Michael O'Hanlon, *The Science of War: Defense Budgeting, Military Technology, Logistics, and Combat Outcomes* (Princeton University Press, 2009), 145.

[2] *Ibid.*, 147. [3] *Ibid.*, 146. [4] *Ibid.*, 147.

[5] *Ibid.*, 151.

would need five to six times as many ships to keep the current force posture in the Persian Gulf and Western Pacific.[6] A major cost-saving measure in recent years has thus been the deployment of ships permanently overseas, allowing the crew to fly back and forth from the United States.[7] This serves to show how the use of combined arms puts a premium on how military operations are conducted. Because of the need to deploy flexible and varied units that fit together in the combined arms paradigm, the units are heavy and thus the costs of deployment are heavy as well.

The heaviness of the US military demonstrates how there is a lot more to a military force operating under the modern system than meets the eye. On television, the soldiers on the ground are all you see, creating expectations among civilians that military forces can be deployed at a moment's notice. This is part of the paradox of military power described in Chapter 2: that the mission that appears swift and decisive depends on an enormous apparatus in which less than half of those employed are actually engaged in combat. This creates a management problem at the core of the modern military system, for the largest efforts in the day-to-day business of the armed forces have very little directly to do with armed forces, being concentrated instead on the much more mundane tasks of administration, training, logistics and so on. For a military leader, staying focused on the ability to deliver decisive military outcomes with armed forces rather than running the military business constitutes a major challenge.

This chapter explains the heavy nature of the US armed forces in order to make general points about the modern military system that provide the business model for the organisation of Western armed forces. The chapter is structured according to the four factors

[6] *Ibid.*, 159. Michael O'Hanlon, *Healing the Wounded Giant: Maintaining Military Preeminence While Cutting the Defense Budget* (Washington, DC: The Brookings Institution, 2013), 37–46.

[7] Michael J. Lostumbo *et al.*, *Overseas Basing of U.S. Military Forces: An Assessment of Relative Costs* (Washington, DC: RAND, 2013), 33–4.

defining the modern military system: (i) combined arms operations, (ii) capital-intensive forces, (iii) focus on 'the tail', that is the logistical and administrative basis of operations, and (iv) flows of information.

COMBINED ARMS

In 1962, Secretary of Defense Robert McNamara divided the US defence budget by activities. For a business analyst who had become CEO, this was a natural thing to do. McNamara wanted to know what the Defense Department 'produced' and how much money the Department was spending producing it.[8] He distinguished between strategic forces, general-purpose forces, mobility forces, the National Guard and reserve forces. Special operations forces were added to the list in 1987. Apart from the force structure, the budget titles list the enabling functions such as intelligence, combat support, administration and so on.

McNamara's production numbers actually measure priorities – if nuclear deterrence was the number-one priority, then strategic forces ought to be allocated the most funds. Over time, the different titles of the US defence budget thus ought to reflect how defence priorities changed in line with changes in the security environment. Or so one would think. As Figure 3.1 shows, the striking fact about the budget by title from 1962 to the present is how little the relationship between the various activities has changed. 'General-purpose forces' is obviously the title that most clearly reflects when the United States are engaged in large operations, and this title thus varies substantially from 1962 until today. General-purpose forces spike during the Vietnam War and after 9/11. However, they also spike in the 1980s during the Reagan build-up of the US armed forces. In other words, the funds spent on general-purpose forces are driven by general trends in the defence budget as well as operational requirements.

[8] Army Force Management School, *Department of Defense Planning, Programming, Budgeting, and Execution (PPBE) Process/Army Planning, Programming, Budgeting, And Execution (PPBE) Process, 2006*, www.acqnotes.com/ Attachments/Army%20PPBE%20Executive%20Primer.pdf (18 October 2014).

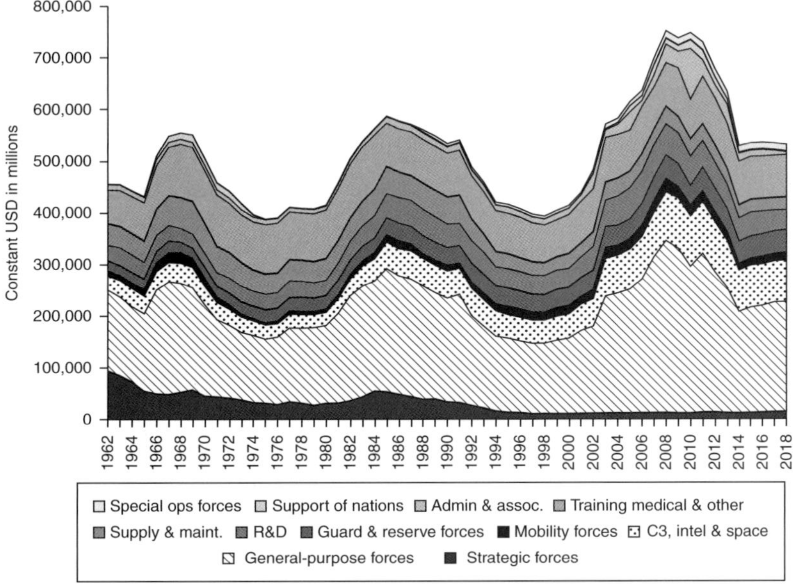

FIGURE 3.1 Department of Defense Total Obligational Authority by programme, 1962–2018 (constant dollars in millions)

This is also clear from the fact that when general-purpose forces are given more funding, so are other titles; and when they are cut, other titles are cut as well. This ebb and flow of the US defence budget is perhaps the most striking feature of Figure 3.1. Substantial changes in priorities can be discerned on only two counts: strategic forces and special operations forces. The investment in strategic (i.e., nuclear) forces has dropped substantially over time, from USD 93 billion when McNamara's measurements were begun in 1962 to a mere USD 11 billion in 2013 (constant dollars). While the end of Cold War nuclear deterrence is thus reflected in the budget, the inclusion in 1987 of special operations forces as a budget title and the increase (from USD 287 million in 1987 to USD 14 billion in 2013, constant dollars) reflects the nature of the post-Cold War missions.[9] Nuclear

[9] National Defense Budget Estimates for FY2014, Department of Defense, Office of the Under Secretary of Defense (Comptroller), May 2013, http://comptroller.defense.gov/ Portals/45/Documents/defbudget/fy2014/FY14_Green_Book.pdf (26 November 2013). Table 6-5.

forces and special operations forces are not the largest items on the budget, even if they count for much in the public imagination – in 1962, strategic forces accounted for 22 per cent of the US defence budget, while the special forces accounted for about 2 per cent of the budget in 2013. The bulk of the US defence budget – then, as now – is spent on conventional forces, and the distribution of funds between these titles has varied very little in fifty years when one takes the massive geopolitical and technological changes in the period into consideration. In 1962, general-purpose forces made up 34 per cent of the defence budget; in 2013, 38 per cent.[10]

This stability is remarkable given the ebb and flow of the US defence budget. When funds have been cut, the cuts have apparently been distributed more or less evenly among the different functions; when funds have increased, the money has been distributed evenly as well. Of course, the titles identify very broad functions and the operations conducted by general-purpose forces have obviously varied considerably in the period. Consequently, one might argue that it is the measurement that is wrong, not the activities. From this perspective, McNamara's titles do not capture the changes in the Defense Department's operations. Even if that were true, however, one would expect variations in how the US military spends its money. If the titles do not reflect operational reality, then a random variation in the numbers over time would be expected. The remarkable consistency in the allocation of funds between activities seems to demonstrate that the bulk of the US defence budget is not really being allocated on the basis of the demands from the outside world but on the internal logic of the Department of Defense. Instead of being a budgetary tool for making the force fit the threats, as McNamara intended, McNamara's budget model seems to demonstrate how the military makes the current operations fit the force.

The remarkable consistency in how the Department of Defense spends its (or rather, the American taxpayers') money would not

---

[10] *Ibid.*, Table 6-4.

surprise Niklas Luhmann. As mentioned in Chapter 2, Luhmann's point is that a system like the modern military system works according to its own logic. Perhaps this is another sense in which the modern military system of combined arms is 'heavy'. The logic running the system is so heavy that its priorities are not easily moved. The balanced force required on the operational level in order to succeed on the modern battlefield is naturally reflected throughout the entire organisation and on every level. This establishes a shared understanding of what the armed forces as an organisation ought to look like; a shared understanding based on the logic internal to the military system rather than on the demands from military solutions from the outside. Furthermore, this way of thinking is reinforced by how the force is supposed to be balanced. One cannot simply change one part of the force, because doing so will have consequences for other parts of the force. This is true on the service level, as it is true between different units and weapon types within services. Problems within the military are thus 'chain-linked'. 'A system has a *chain-link* logic', Richard Rumlet explains, 'when its performance is limited by its weakest subunit, or "link".'[11] This linkage means that a change to the organisation will get nowhere by focusing on the parts. It is the entire system, and especially the linkages between the elements of the system, which one ought to focus on if one wants to change anything. Linkages are often guided by conceptions about what is needed to get a certain job done. During the 1990s, for example, two US Army divisions were labelled 'unready' because each of them had sent a brigade to the Balkans; but they still had two brigades each. Two brigades are heavy – they use 1,200 tons of supplies per day in combat; and most European armies would happily define two such brigades as a division. In the context of the US Army, however, the deployment of one brigade created a linkage issue. Using a third of the division's resources tied up the entire division. Consequently, the

---

[11] Richard Rumlet, *Good Strategy/Bad Strategy: The Difference and Why It Matters* (London: Profile Books, 2011), 116.

cost of deploying one brigade was – at least in terms of activity – the same as the deployment of a division. Obviously, the solution to this vast discrepancy between cost and effort should result in an organisational change, but the entire divisional structure reflected the needs of the modern military system to maintain a balanced, combined arms organisation.[12] The operations in Iraq and Afghanistan now highlight these problems and give rise to the COIN approach to reforming the US armed forces (more on this in Chapter 4). Understanding the chain-linked nature of modern military units requires closer examination of the so-called tooth-to-tail ratio.

TOOTH-TO-TAIL

The 'tooth-to-tail' concept describes 'the number of troops in a military organization employed in combat duties versus the number functioning in noncombat roles', John J. McGrath explains.[13] McGrath works at the Combat Studies Institute, Fort Leavenworth, Kansas, and his ground-breaking study of tooth-to-tail ratios carefully examines the nature and growth of the US Army's tail, from the creation of the modern system during the First World War to the Iraq War. McGrath focuses on expeditionary warfare, which is important because doing so makes his study a survey of how the Army is used rather than how it is organised.[14] McGrath carefully points out that noncombat roles are not just logistics, but also administration, public relations, people who runs camps, military bands and the vast array of other functions needed to make the modern military system work.[15]

Perhaps the most striking feature of the modern military system is the disproportion between the number of actual combat soldiers and the number of military personnel employed in 'back-office'

<hr>

[12] O'Hanlon, *The Science of War*, 35.

[13] John J. McGrath, *The Other End of the Spear: The Tooth-to-Tail Ratio (T3R) in Modern Military Operations* (Fort Leavenworth, KS: Combat Studies Institute Press, 2007), 2.

[14] *Ibid.*, 4.   [15] *Ibid.*, 5–8.

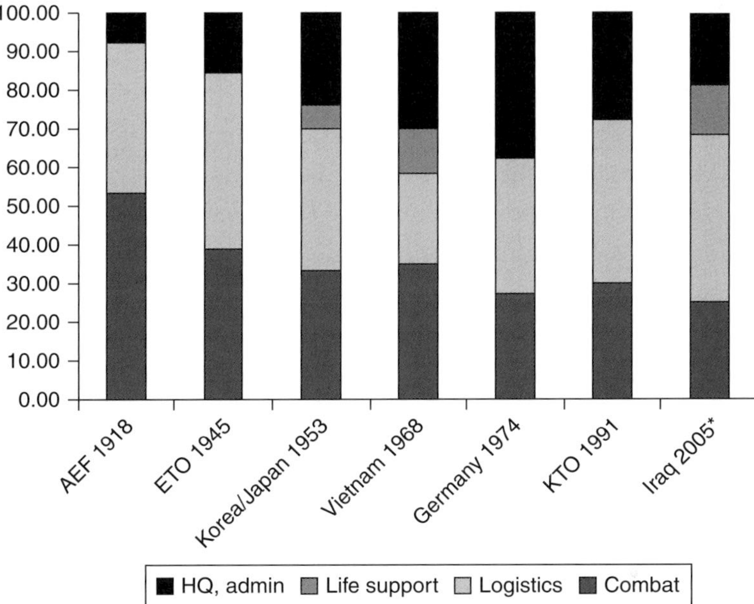

FIGURE 3.2 Distribution of functions in US Army contingents in theatre, 1918–2005
* including contractors

functions. An infantry division in the American Expeditionary Force in France during the First World War employed 7 per cent of its strength in headquarters, whereas 24 per cent of the soldiers allocated to a US Army brigade in Iraq in 2004 were allocated to administrative tasks.[16] Figure 3.2 illustrates the distribution of functions in US Army contingents in the largest theatres of operation since the First World War.

Between 1918 and 2005, the ratio of combat troops to administrative staff in a given theatre of operations has fallen steadily. It is interesting to note, however, that the ratio of troops assigned to logistical functions has remained rather constant – about one-third of the troops deployed. It is the number of soldiers behind a desk performing various administrative tasks that has increased dramatically – from

---

[16] *Ibid.*, 102.

8 per cent in 1918 to 38 per cent in the US Army in Germany in 1974. Since then, the Army has worked hard to reduce this figure, which thus dropped to 28 per cent in the first Iraq War (1991) and 24 per cent in the second (2003–11). McGrath even argues that if one takes contractors and personnel in Kuwait into account, only 18 per cent of the troops in Kuwait were fighting the war from behind a desk. That is the lowest number of personnel employed in headquarters and administrative work since the Second World War. One reason for this is that the US Army has been able to reduce the size of the operational unit, thereby reducing the demand for coordination and planning at the top. This also meant that the bulk of the army was not deployed in theatre. Thus, one might argue that the reason for the low number of headquarters personnel in Iraq was actually a reflection of the fact that the bulk of the Army, including its administration, was not deployed. In order to investigate this further one must examine the operational units more closely.

Between 1918 and 2005, the size of the operational unit in the US Army fell. The operational unit of the US Army in the First World War was a division of 28,105 soldiers. In Iraq and Afghanistan, the Army deployed in modular combined arms brigades of 3,735 soldiers.[17] McGrath has taken a look inside these units to study the tooth-to-tail ratio in the operational unit. Whereas Figure 3.2 shows the tooth-to-tail ratio of the entire force deployed in a given theatre, Figure 3.3 shows the tooth-to-tail ratio for the units deployed to the front line.

'The complexity of modern weapons requires increased numbers of trained personnel farther to the rear for their maintenance', General Maxwell Taylor observed in 1959. This tendency had begun before the General made this remark and would continue afterwards.[18] During the First World War, an AEF infantry division had a 2:1 ratio between tooth and tail.[19] During the Second World War,

---

[17] *Ibid.*, 102.

[18] Maxwell D. Taylor, *The Uncertain Trumpet* (London: Atlantic Books, 1959), 174.

[19] McGrath, *The Other End of the Spear*, 11.

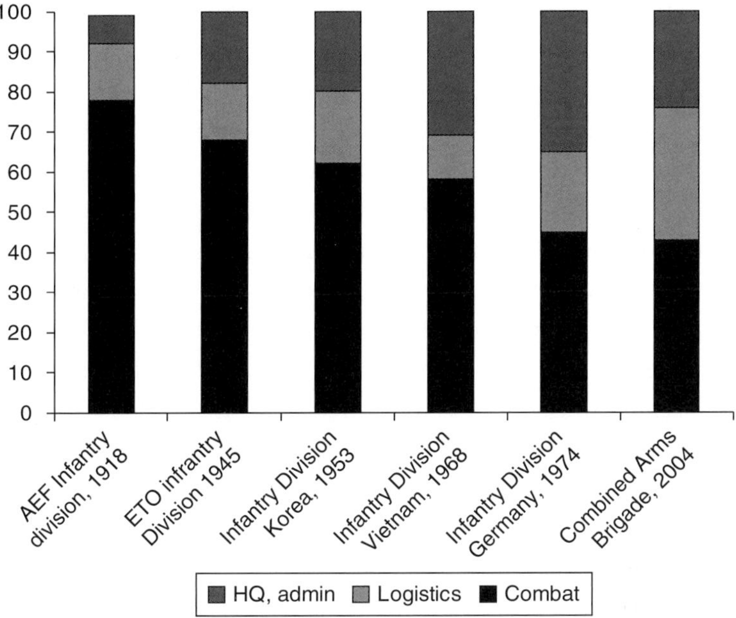

FIGURE 3.3 Distribution of functions in operational units of the US Army, 1918–2004

HQ staff rose to 18 per cent despite the Army's best efforts to shorten the tail.[20] These efforts were as ineffective during the Second World War as they were in Korea, where the administrative function in an infantry division was up by two percentage points. In 1968, a US army infantry division in Vietnam was down to 58 per cent of its soldiers being deployed in combat. After the Vietnam War, the number of combat soldiers fell to under 50 per cent in an infantry division in Germany in 1974. This was clearly unacceptable, and the Army tried to make itself more effective in the 1980s by a concept known as Division 86 – Army of Excellence. Division 86 was built around new platforms – the M1 tank, the M2 infantry-fighting vehicle and Apache attack helicopters. This new military force was deployed

[20] *Ibid.*, 102.

in Operation Desert Storm in 1991.[21] Still, the trend of making the Army heavier in the back and thinner in front seemed unstoppable. In 2004, the Army was able to reverse the trend by changing the organisational unit from the division to the brigade. A heavy division had 39 per cent of its troops in combat roles, a combined arms brigade 43 per cent in combat roles, while a Stryker brigade had 52 per cent of its troops in combat roles.[22] This made the Stryker brigade the most effective unit in terms of the tooth-to-tail ratio in the US military since the 1960s.[23] From 2002, the Army used the Stryker brigade as the overall template for the construction of a 'modular army'. The new divisions were 5,000 troops larger than the previous design, as deployed in Kuwait, and HQ elements made up 56 per cent of that increase, with most of the rest in logistics, while the modular brigades retained a higher tooth-to-tail ratio – 1:1.3 as opposed to 1:1.02 in the previous configuration.[24]

During the 1990s and the first decade of the twenty-first century, the Army worked to optimise the tooth-to-tail ratio. As had been the case during the Second World War, this was largely achieved by removing functions from the organisational charts rather than actually removing the need for these services. Thus, if only counting the uniformed personnel, the US Army had a 1:2.5 tooth-to-tail ratio in Iraq in 2005, with 40 per cent of the soldiers in combat functions. In other words, the army was as effective in tooth-to-tail terms as the divisions deployed in Europe during the Second World War. However, McGrath finds these numbers to be deceptive. He includes civilian contractors in the equation, which reduces the combat percentage to 25 per cent – the lowest level in any US Army operation since 1917.[25]

With only 25 per cent of the people deployed in combat operations and only 11 per cent of the Army deployed to Iraq in the first place, the personnel actually in combat risk feeling alienated from

---

[21] Keith L. Shimko, *The Iraq Wars and America's Military Revolution* (Cambridge University Press, 2010), 26–90.

[22] McGrath, *The Other End of the Spear*, 102.

[23] *Ibid.*, 44–50.   [24] *Ibid.*, 46–7.   [25] *Ibid.*, 50–5.

the military organisation as such. Why are they in harm's way when the rest of the Army is working safely on a base? Colonel Burton, commander of Dagger Brigade Combat Team, 1st infantry division, in Iraq, expressed such sentiment in a memo that received widespread attention when published on the homepage of the *Small Wars Journal*. Colonel Burton argued that the difference between the conditions for the officers who were deployed and those who were not was too great. 'They see that they represent a small portion of the Army's Officer Corps that has actually deployed to fight this war', Colonel Burton wrote of the officers in his command, 'and they are frustrated by the knowledge that the majority of their peers may not have deployed thus far. They have spent the past four years in a continuous cycle of fighting, training, deploying, fighting, etc. and they see no end in sight.'[26] The Pentagon clearly picked up on this. In 2011, then-Secretary of Defense Robert Gates addressed cadets at West Point about the officers at the front who are schooled in taking risks and responsibility in the battlefield, but Secretary Gates then noted:

> The opposite is too often true in the rear-echelon headquarters and stateside bureaucracies in which so many of our mid-level officers are warehoused. Men and women … may find themselves in a cube all day re-formatting power point slides, preparing quarterly training briefs, or assigned an ever expanding array of clerical duties. The consequences of this terrify me.[27]

Secretary Gates was sensitive to the human resource challenge which the tooth-to-tail ratio represented, but the striking fact about his remarks is that for all his lamenting of the current state of affairs,

[26] J. B. Burton, 'The Officer Critical Skills Retention Bonus', *Small Wars Journal*, 18 July 2007, http://smallwarsjournal.com/blog/the-officer-critical-skills-retention-bonus (24 January 2012).

[27] Robert M. Gates, Speech at United States Military Academy (West Point, NY). As Delivered by Secretary of Defense Robert M. Gates, West Point, NY, 25 February 2011. Defense.gov Speech, www.defense.gov/speeches/speech.aspx?speechid=1539 (24 January 2012), §23.

the Secretary of Defense seems powerless to do anything about these problems. The fact that the Secretary of Defense seemed compelled to point to a problem rather than a solution speaks volumes about how entrenched the uneven tooth-to-tail ratio has become. The Defense Department's commitment to 'rebalancing tooth and tail' in QDR 2014 should be read in that context.[28]

In terms of operational effectiveness, a larger tail does not necessarily translate into a lower capacity to fight, however; the point of the modern military system is in many ways the opposite. It is worth noting that the main increase in the tail has been in headquarters and administrative staffs rather than in logistics. Running the Army is thus consuming increasing amounts of resources – in the theatre as well as in combat units. It seems as though the modern system itself requires more and more resources to run. This increase in the administrative burden of the system stands in sharp contrast to the dramatic increase in the effectiveness of individual units. The main unit of operation has decreased by 752 per cent. This serves to show the staggering impact of military technology that has enabled a Modular Combined Arms Brigade of 4,735 soldiers to become the unit of choice rather than the 28,105-soldier-strong infantry division used by the American Expeditionary Force in 1918. These numbers are of course Army numbers, and the Air Force and Navy (in the United States and elsewhere) are organised differently. Yet because the Army employs so much more manpower, Army numbers can highlight trends which are perhaps more difficult to discern in other services. The modern system has become increasingly capital-intensive, which is the subject of the next section.

CAPITAL-INTENSIVE FORCES

'In the year 2054, the entire defence budget will purchase just one aircraft', Norman Augustine wrote, 'this aircraft will have to be shared by the Air Force and Navy 3½ days each per week except for leap

---

[28] *Quadrennial Defense Review 2014*, xi.

year, when it will be made available to the Marines for the extra day.'[29] Having served as Assistant Director of Defense Research and Engineering in the Office of the Secretary of Defense as well as President and CEO of Lockheed Martin, Augustine knew the demand side of the defence establishment as well as the supply side. He was able to formulate in an aphorism what people with inexperience in either industry or the military perhaps grasped less clearly. Secretary of Defense Robert Gates conceded this point in a 2009 article in *Foreign Affairs*:

> When it comes to procurement, for the better part of five decades, the trend has gone toward lower numbers as technology gains have made each system more capable. In recent years, these platforms have grown ever more baroque, have become ever more costly, are taking longer to build, and are being fielded in ever-dwindling quantities. Given that resources are not unlimited, the dynamic of exchanging numbers for capability is perhaps reaching a point of diminishing returns.[30]

Compared to 1970, the US defence budget had risen to index 140 in 2010, while the number of platforms bought for the government's dollars had fallen to around index 40 in terms of warships and fighter aircraft.[31] In short, the United States spends more on fewer platforms that are engaged in more costly operations with the participation of fewer soldiers, airmen and marines. By way of an analysis of how the Department of Defense has spent its budget from 1948 to 2013 (the analysis is predominately based on the Department of Defense's Budget Estimates for Financial Year 2014 and thus includes projections for spending up to 2018), this section describes the reality behind Augustine's aphorism.

---

[29] Norman R. Augustine, *Augustine's Laws* (New York: Viking, 1986), 111.

[30] Robert M. Gates, 'A Balanced Strategy: Reprogramming the Pentagon for a New Age', *Foreign Affairs*, January (2009), 5.

[31] 'The Cost of Weapons: Defence Spending in a Time of Austerity', *The Economist*, 26 August 2010, www.economist.com/node/16886851 (6 January 2012).

The long-term trends in the US defence budget have been increased spending on operations and maintenance as well as procurement. From 1960 to 2013, Department of Defense personnel spending increased on average 0.2 per cent annually. This number obviously hides massive variations – from a 10.3 per cent decrease from 1992 to 1993 to a 27.8 per cent increase in spending on personnel from 1984 to 1985.[32] These considerable variations in personnel costs possibly reflect the fact that personnel is a more variable resource than, say, fighter aircraft, because soldiers can be recruited and trained faster than an aircraft can be procured. Still, the Department of Defense spends much less on military pay today than in 1960. Accounting for inflation, which accounts for much of the growth in the nominal expenses to personnel, military pay accounted for USD 137,773 million in 1960, while the Department of Defense spent USD 77,182 million on military pay in 2013. This is reflected in the fact that the Department of Defense has cut its military manpower by one million people from 1960 to 2013 – from USD 2.5 million in 1960 to USD 1.5 million in 2013.[33] This is one reason why the operative unit of the army is no longer a division but rather a brigade combat team. It is worth noting, however, that the aggregated costs for personnel have hardly changed. Separating inflation from the numbers, the US military spends almost the same amount on personnel in 2013 as in 1960 (in constant dollars, 156,390 million as opposed to 150,817 million).[34] The costs associated with the all-volunteer force of today are much different from those of the conscript force of the 1960s, however. The costs of military health care, pensions and other military compensations have increased consistently, and above inflation. The Congressional Budget Office thus estimates that the real growth of military health care will be 4.6 per cent per year on average in the coming decade.[35] These costs are funded by the accounts

[32] *National Defense Budget Estimates for FY 2014*, Table 6-1.

[33] *Ibid.*, Tables 6-2 and 7-5.   [34] *Ibid.*, Table 6-1.

[35] Congressional Budget Office, *Costs of Military Pay and Benefits in the Defense Budget* (Washington, DC: Congress of the United States, 2013), 3.

for personnel as well as operations and maintenance,[36] and personnel costs are thus a driver behind several titles of the US defence budget.

One measure of how capital-intensive the US military has become is to compare how much of the budget is 'spent' on one soldier, airman or marine. Measuring the capital-intensiveness of the US armed forces in terms of expenditures per serviceperson, the Department of Defense spent USD 160,677 per soldier, airman or marine in 1960, whereas it spent USD 431,249 in 2013.[37] In 1960, 114,672 constant dollars were spent per person employed in the Army, in 2013 USD 364,397 were spent per person employed in the Army. The costs per person in the Army had thus tripled, while the per-capita costs in the Navy and Air Force had doubled from 1960 to 2013.[38] Since the costs for military pay dropped by USD 60,591 million (constant) in the same period (1960–2013), this increase is obviously not due to service personnel being paid more, even if US service personnel are generally being paid well compared to others with a similar background in civilian jobs.[39] On the contrary, the soldier–budget ratio demonstrates that the military's main production effort is not being delivered by humans, but rather by machines. It is the materiel and costs of deploying and maintaining it that is in the budget. Figure 3.4 shows the relationship between expenditures on personnel, operations and maintenance and procurement and research and development.

Figure 3.4 demonstrates how the personnel and operations costs are roughly similar from 1948 to the mid 1970s. Procurement is aligned with the two other main accounts, but increases in wartime. There are thus lulls in procurement after the Korean (1950–3) and Vietnam Wars (1965–73), but it quickly picks up again and becomes the larger part of the budget in the early 1960s and from the mid 1970s. The Reagan defence build-up results in a procurement spike in the 1980s, followed by a lull in the 1990s and a build-up after 2001 that culminates in 2008, after which the budget is significantly reduced.

---

[36] *Ibid.*   [37] *National Defense Budget Estimates for FY 2014*, Tables 6-8 and 7-5.

[38] *Ibid.*, Tables 6-13 and 7-5.

[39] O'Hanlon, *Healing the Wounded Giant*, 62.

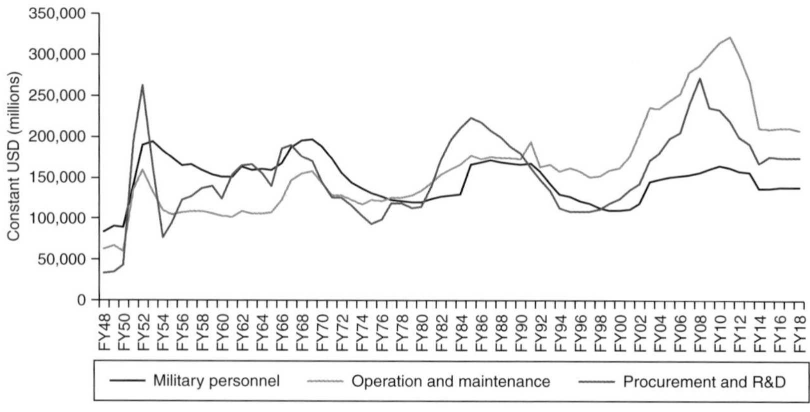

FIGURE 3.4 US defence expenditure by title, 1948–2018 (constant USD in millions)

Procurement follows a set pattern of investments, a pause and new investments. Pay remains the lowest cost of the three. 'Operations and maintenance' is a title that obviously follows the deployment of US forces. Yet, Figure 3.4 allows us to take the long-term view, and in those terms it is clear that while operations and maintenance vary considerably with operational needs, these needs seem to be continuously increasing. This culminates in 1991, when operations and maintenance surpasses procurement as the most expensive item on the budget. This corresponds with McGrath's findings that the tail consumes a greater share of the Army's resources. This change in how the Department of Defense spends its money is perhaps best appreciated against the annual growth in personnel, procurement and R&D as well as spending on operations and maintenance. Since we know from Figure 3.4 that the Korean and Vietnam Wars created huge variations in procurement growth, Figure 3.5 presents the annual growth from 1960 to 2010.

From 1960 to 2013, the average annual growth in R&D and procurement was 3 per cent per year.[40] Here again, the variation is

---

[40] 2013 numbers are from the 2014 defence budget. *National Defense Budget Estimates for FY 2014.*

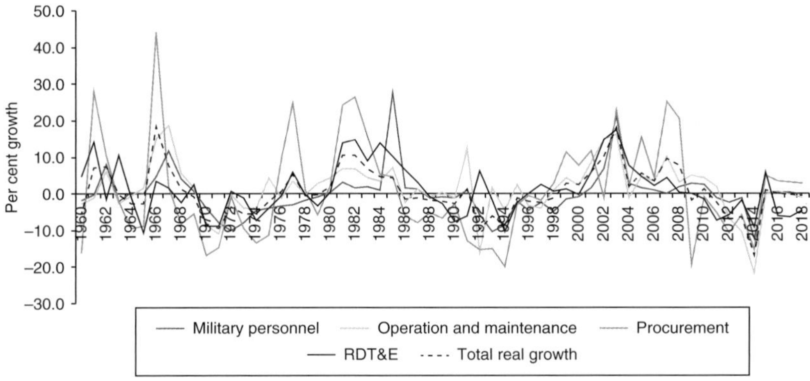

FIGURE 3.5 Yearly real growth of US defence budget, 1960–2018

considerable – the numbers include a drop in procurement expenditure of 15 per cent in 1993 and a 23 per cent increase in 2003. These considerable variations average out the fact that spending on procurement and R&D increased from USD 124,101 million in 1960 to USD 271,513 million in 2008 and is expected to fall to USD 174,336 million in 2018 (constant dollars). In 1960, operations and maintenance had increased to USD 103,872 million. The 2014 budget puts operations and maintenance at USD 210,200 million, down from 268,217 million in 2013 (constant dollars). Obviously, this budget title varies with operations. The 2009–15 numbers reflect the drawdowns of the wars in Afghanistan and Iraq. Thus, the operational and maintenance costs in 1999 were only USD 159,218 million. From 1960 to 2013, operations and maintenance had an average annual growth of 1.9 per cent. This increase has been so steady, however, that operations and maintenance costs have increased by 158 per cent.[41] This increase is associated with the fact described by Augustine that the United States is spending more and more money on fewer and fewer platforms. This obviously increases the operational tempo for an individual platform and its crew, which is also why the Pentagon planners hope to reap a substantial windfall

[41] *Ibid.*, Table 6-1.

from the return of the American troops from Iraq and Afghanistan. In 2009, operations and maintenance costs thus fell by 19.3 per cent following the withdrawal from Iraq, and these costs were expected to fall by 21.6 per cent in the budget for 2014.[42]

These numbers and tables demonstrate that Augustine was not merely being funny when he predicted that the services would have to find some way to share a single airplane if the US defence budget continues to be spent at the current rate. Augustine's aphorism describes an unsustainable situation, one in which the capital-intensiveness of the modern military system undermines the utility of the military organisation as such. In 2040, what exactly are the more than 300,000 people working in the US Air Force to spend their time on when they are not flying? The answer, of course, is that they will find ways to keep themselves busy, and thus Augustine points out the inherent waste in a system where the tooth-to-tail ratio becomes so lopsided that one would expect the organisation to be able to work just fine without operational platforms. Some will argue that Augustine overlooks a crucial fact, however; it is not the number of units that matters, but the military effect of the units deployed. If one plane can deliver the payload needed, then one plane is all you need.

General David Deptula of the US Air Force makes this effect-based argument: 'in some cases, a single aircraft and one PGM [precision-guided munition] during the Gulf War achieved the same result as a 1,000-plane raid with over 9,000 bombs in World War II – without the associated collateral damage'.[43] General Deptula was one of the officers planning the Gulf War air campaign and became a chief apologist for effect-based operations planning. Where the generals of the First World War eventually realised that a modern battle was won by concentrating and coordinating arms at a decisive point in order to achieve a breakthrough, the Gulf War Air Force planners were

---

[42] *Ibid.*, Table 6-1.

[43] David A. Deptula, *Effects-Based Operations: Change in the Nature of Warfare,* February 2001, Aerospace Education Foundation, www.airforce-magazine.com/ SiteCollectionDocuments/TheDocumentFile/Strategy%20and%20Concepts/ EBO_deptula_2001.pdf (6 January 2012), 9.

much more ambitious. They believed that airpower made it possible to ignore the front line on the Saudi–Kuwait border. They wanted to fly over the Iraqi positions and use PGMs to take out specific nodes in the Iraqi military network, thus rendering it impossible for Iraqi dictator Saddam Hussein to control his forces on the front line. They focused on the transmitters of information in the system and devised a strategy to take them out. Even if the air campaign was not as successful as some reports during the war led the public to believe, it was still an impressive achievement considering the number of sorties in relation to the number of targets hit. The 1991 operations in Iraq thus provided a blueprint for a more air-dependent American form of warfare.[44]

The success of US airpower seems to suggest that Augustine overlooks the fact that while the US Air Force and Navy have fewer and more expensive planes, these planes are also vastly more effective than the platforms of the 1970s. Perhaps it is the relationship between price and effectiveness that should be measured rather than the ratio of price to number of units? In fact, this argument demonstrates how the modern system has led to a focus on platforms rather than on the force structure in general or the development of said force structure. A single plane might be enough for a particular commander for a particular operation; for the President or Defense Secretary, however, who must develop a balanced force for global use, the number of platforms in relation to the number of missions does matter. If a plane is bombing in Afghanistan, it cannot simultaneously bomb in North Korea. This is the price to pay in terms of strategic agility for fewer and more expensive platforms. In organisational terms, the price is the development of an increasingly tail-heavy organisation. One reason why the operations and maintenance costs have risen is the need to ensure a quick turn-around

---

[44] John Warden, 'Air Theory for the Twenty-first Century', in Barry R. Schneider and Lawrence E. Grinter (eds.), *Battlefield of the Future: 21st Century Warfare Issues*, The Air War College Studies in National Security (Maxwell Air Force Base, AL: Air War College, 1995).

for a more limited number of technologically complicated platforms. Fewer platforms also increases the need for planning in order to use the available platforms optimally. In fact, these consequences of the capital-intensiveness of the modern military system reinforce the focus on operational effectiveness, which leads General Deptula and others to disregard the consequences of operationally effective planes on the force structure in general. Cost becomes a secondary consideration when focusing on military effect alone.

Revealingly, Deptula compares the effectiveness of 1990s planes with 1940s planes. Instead of comparing a B-2 with a B-17, Deptula could have compared an F-22 with an F-16. An F-22 costs USD 150 billion (2009 level) per unit, whereas an F-16 costs USD 32 billion.[45] Had Deptula made this comparison, the marginal operational effectiveness achieved per million dollars paid for the new plane would have been much lower. This type of argument is often between old and bad versus new and good rather than between middle-aged and good enough. The underlying argument is that a more effective military operation merits the increased costs, even if a similar effect could have been achieved using a cheaper platform. Niklas Luhmann points out how organisations often seek to reduce their own risk by transferring risks to others.[46] This is what the US Air Force is doing in this case. Quite understandably, Air Force planners like General Deptula are focusing on maximising operational effect and minimising operational risk. The lower the risk of failure in terms of planes being shot down or pilots missing their targets, the more effective campaign plans can be made. Minimising operational risk does not make risk disappear, however. Instead, operational risk is being translated into an 'organisational risk' by creating an increasingly capital-intensive and expensive military production line.

A capital-intensive military constitutes a risk in two ways: (i) when operational effectiveness is correlated with technological

---

[45] O'Hanlon, *The Science of War*, 28.

[46] Niklas Luhmann, *Risk: A Sociological Theory*, transl. Rhodes Barrett (New York: Walter de Gruyter, 1993).

innovation, then the armed forces are compelled to keep investing ever more heavily in materiel; (ii) in turn, this places an increasing burden on government budgets. This risk is associated with the modern military system and not unique to the United States as such. And in that fact rests the genuine military strength of the United States, a military strength that is not measured in battles lost or won, but rather in dollars. An opponent, fighting within the framework of the modern military system, bears the risk of being able to innovate their military technology as well as paying for it. The US Army Concepts Analysis Agency has computed data on historical campaigns and concluded that when military capabilities on either side were roughly equal, the attacker won 58 per cent of the 230 battles recorded.[47] In other words, if you are balanced with your opponent, you had better attack and be a better soldier. However, the data suggests that the safest way to ensure victory is to make sure that attacker and defender are not equal in military capability. If you want to improve your odds, you have to achieve superiority – provided you can pay for it. By paying for a military budget greater than the budget of the nine largest military budgets combined, the United States is improving its odds in any future conflict. This operational risk-reduction has structural consequences as well.

Big budgets are grand strategy. Constantinos Markides notes that top-ranked firms in a particular industry have a 91 per cent probability of survival.[48] 'Without the benefit of a technological innovation', Markides argues, 'it is extremely difficult for any firm to successfully attack bigger competitors or to successfully enter new markets where big established players rule.'[49] From this perspective, the size of the US defence budget is a point in and of itself. This approach to the defence budget was institutionalised in the

---

[47] O'Hanlon, *The Science of War*, 64.

[48] Constantinos C. Markides, *Game-Changing Strategies: How to Create New Market Space in Established Industries by Breaking the Rules* (San Francisco, CA: Jossey-Bass, 2008), ix.

[49] *Ibid.*, xii.

1970s and took its point of departure in a memo entitled *Strategy for Competing With the Soviets in the Military Sector of the Continuing Political–Military Competition* written to the Secretary of Defense by Andrew Marshall, who has been the Director of Net Assessment in the Office of the Secretary of Defense since 1973.[50] The most obvious and most difficult question in strategy is 'What do we want to achieve?', followed by 'Which resources do we have to achieve that goal?'. Marshall saw more clearly than most how the two questions were connected in US defence policy in the mid 1970s, a time when investments in defence were slumping in the wake of Vietnam (see Figure 3.5). The Vietnam War no longer drove the budget in terms of operational needs, and the debacle in Vietnam meant that the US armed forces were looking for other templates for operations as the basis for the future force structure.

From Marshall's perspective, this was the wrong approach. In strategy terms, the military budget was not about operations, Marshall argued, but about risk management. In the memo, Marshall notes that 'current defense programs carry with them an implicit strategy this necessities planning at least the same time frame as procurement – 30 years'.[51] Writing in another time of austerity, Marshall notes that 'we can no longer indulge in the "rich man" strategy of insuring against all possible adverse futures'.[52] Instead, Marshall asks the Secretary of Defense to identify what really matters: the conflict with the Soviet Union. The United States must focus on long-term competition with the USSR and 'actively seek out Soviet weaknesses and vulnerabilities which can be exploited by changes in US doctrine, tactics and/or forces'.[53] In other words, the US strength was the size of the defence budget; and the Soviet Union's weakness was the size of its defence budget, because the United States could afford

---

[50] Andrew Marshall, *Strategy for Competing With the Soviets in the Military Sector of the Continuing Political-Military Competition*, Department of Defense Office of Net Assessment, 1976, mimeo, http://goodbadstrategy.com/wp-content/downloads/StrategyforCompetingwithUSSR.pdf (20 January 2012).

[51] *Ibid.*, 3.   [52] *Ibid.*, 7.   [53] *Ibid.*, 26.

a massive defence budget, whereas the Soviet Union could not. The United States should thus use existing technologies more effectively and develop new technologies more quickly than the Soviet Union in order to 'cause Soviet systems to be obsolete', thus 'eliminating the returns of previous Soviet investments'. So doing, the United States would be able 'to control the pace of competition'.[54]

Even if the military embarked on AirLand Battle in a pursuit of a new operational concept to shape investments against Marshall's recommendation in the memo, Marshall's overall strategy was largely implemented. The US defence budget had fallen from USD 518 billion in 1968 to 314 billion in 1976 (2005 constant dollars). In 1979, a gradual increase in the funds allocated to national defence began, until the spending spree peaked at USD 482 billion in 1989. The point is that the US defence budget has never returned to 1970s levels. The Reagan investments seem rather modest compared to the USD 548 billion the Department of Defense spent in 2008. And even if that number is supposed to drop to USD 455 billion in 2018, that still represents a larger national investment in defence than the US government spent from 1971 to 2003 (expect for 1986–90) – and one should perhaps note that the United States never spent more than USD 400 billion on national defence from 1955 to 1967.[55]

Marshall's strategy was spectacularly successful. Instead of using R&D and the budget to fulfil strategic objectives, Marshall's memo makes the budget a strategic tool in its own right. Perhaps one of the most perceptive elements of it was that – schooled in historical materialism – the Soviet planners thought in exactly the same terms as Marshall. They were thus prepared to concede defeat when they realised that their industrial 'base' could not compete with the

[54] *Ibid.*, 27.

[55] The reference here is to what is appropriated to national defence, excluding veterans, space, etc. *National Defense Budget Estimates for FY 2014*, Department of Defense, Office of the Under Secretary of Defense (Comptroller), May 2013, http://comptroller.defense.gov/Portals/45/Documents/defbudget/fy2014/FY14_Green_Book.pdf (26 November 2013). Table 7-2.

United States.[56] The defence industry lobby and all the other groups with an interest in a large US defence budget were equally prepared for Marshall's argument, which in fact provides a rationale for using tax dollars to pay for ever more advanced military platforms without undue concern about either cost or usability. It was a wise grand strategy that made the military–industrial complex happy. This ensured the constituency for actually implementing the strategy, which is always one of the weak points in any grand strategy.

In his book *Good Strategy/Bad Strategy*, Richard Rumlet mentions Marshall's paper as an example of a good strategy. Marshall and co-author James Roche did what Rumlet believes strategists should do: 'identify your strengths and weaknesses, assess the opportunities and risks (your opponent's strengths and weaknesses), and build on your strengths'.[57] A strategy that is so successful and has powerful backing from vested interests is very difficult to change. The fact that the Soviet Union was defeated did not make the United States deviate from the path set out by Marshall. Figure 3.5 shows the drop in defence growth in the 1990s, but Marshall reinforced his argument at the time by introducing the notion of revolution in military affairs (RMA), pointing out that the technological achievements financed by his strategy constituted a revolution in military affairs – a revolution the USA will have to dominate.[58] In the 1970s, the US military budget was to outpace that of the Soviet Union, in the 1990s the US military budget was to compete with itself and the innovations it produced in the 1970s and 1980s. As Markides notes, 'an innovation is considered strategic if it is difficult for competitors to imitate, substitute, or replicate quickly'.[59] It was this innovation high ground Marshall wanted to secure for the United States. The military leadership embraced the

---

[56] Mikkel Vedby Rasmussen, *The Risk Society at War* (Cambridge University Press, 2006), 45–52.

[57] Rumlet, *Good Strategy/Bad Strategy*, 28–31.

[58] Andrew W. Marshall, *Some Thoughts on Military Revolutions*, Second Version, Office of the Secretary of Defense, Director of Net Assessment, Washington, DC (1993).

[59] Markides, *Game-Changing Strategies*, 18.

RMA by the mid 1990s. In 1997, the Chairman of the Joint Chiefs, General John Shalikashvili, described the RMA as the reference point for future strategy in much the same terms as Marshall had described the Soviet Union in 1976: 'warfare is changing with the growth of technological change, and we must not only stay abreast of it, but dominate it'.[60] This was a reason to keep technological innovation going. As Carl Conetta notes,[61] the large investments in materiel in the Reagan years meant that the US military was well-stocked and able to invest in RMA technologies despite the fact that the national defence's share of US GDP fell from 5.6 per cent in 1989 to 3 per cent 1999 (see Figure 6.1).

The modern military system is capital-intensive. This has produced a force structure that costs more and more while fielding fewer and fewer platforms. With a defence budget larger than the nine biggest defence spenders in the world, the United States probably fields the most capital-intensive forces in the world. This is why it makes sense to study the modern system from an American perspective. And from an American perspective, the ability to outspend and outpace potential competitors in terms of cost and innovation has been a strategic objective in itself from the 1970s. In spite of the fact that the US military appropriates less of the American national wealth than it did during the Cold War, the money spent on procurement and research and development has thus increased dramatically in real terms. This is the result of a focus on machines rather than men. In order for these machines and the men that operate them to be able to operate, the exchange of information is central. This is basically the essence of the RMA. The next section will deal with the fourth element of the modern military system: information.

---

[60] *Quadrennial Defense Review Report*, May 1997 (Washington, DC: Department of Defense, 1997), Preface.

[61] Carl Conetta, *An Undisciplined Defense Understanding the $2 Trillion Surge in US Defense Spending*, Project on Defense Alternatives, PDA Briefing Report #20, 18 January 2010, www.comw.org/pda/fulltext/1001PDABR20.pdf (2 January 2013), 10–11.

INFORMATION

One day in 2010, US Army Colonel Lawrence Sellin had finally had it with his job at International Joint Command, ISAF, in Afghanistan. 'Like most military organizations', he wrote of the Command, 'structure always trumps function'.[62] Colonel Sellin felt a complete lack of connection between the management of the war at the headquarters and the actual fighting outside the headquarters. The war was described on PowerPoint slides rather than experienced in the field. Sellin thus felt that the work in the headquarters made little difference to how the war was fought. He described how he and his fellow officers prepared the slides each day for the CUA (Commander's Update Assessment), in spite of the fact that the commanders seldom bothered to attend this briefing about what had happened in the theatre in the previous twelve hours. Sellin sarcastically notes that 'the CUA slides only change when a new commander arrives or the war ends'.[63] Whether the war ends or not actually seemed a secondary consideration to the briefing itself. 'The ability to brief well is, therefore, a critical skill', Sellin concludes. 'It is important to note that skill in briefing resides in how you say it. It doesn't matter so much what you say or even if you are speaking Klingon.'[64]

Ultimately, what Sellin said did matter. Shortly after the wire service UPI published his complaints, he was fired; not for what he said, the Army was careful to point out, but because he had talked to reporters without clearing it through official channels.[65] Colonel Sellin probably did not appreciate the irony, but the way he was fired only reinforced his argument that the Army was concerned with procedure rather than results. This would hardly surprise a systems theorist, however. According to Luhmann, a system is not defined in terms of function but in terms of its own complexity. From a systemic perspective, the purpose of the ISAF headquarters in which Colonel Sellin worked was not to win the war but to manage the

---

[62] 'The PowerPoint Rant that got a Colonel Fired', *Army Times*, 2 September 2010.
[63] *Ibid.*   [64] *Ibid.*   [65] *Ibid.*

system. Tons of materiel and thousands of people from forty-six different nations, with their particular political agendas, military materiel and operational preferences, had deployed to Afghanistan. This was a complex operation to manage even if a single soldier never left camp. The fact that soldiers had to fight the Taliban only complicated the operation. Sellin's feeling of disconnect between managing the operation and conducting the mission thus demonstrates a crucial point about information and coordination in the modern military system.

The modern military system replaces leadership in battle with battle management. One might define leadership as the use of personal competences, skills and knowledge to give authoritative direction to a group of people or an organisation. Where leadership is personal, management is the processes, decisions and leadership actions that make the planning, execution and evaluation of operations possible. The US Army divisions in the First World War needed comparatively few staff officers, because generals in battle relied on their subordinate commanders to lead the soldiers in the field. With the introduction of the modern system, an infantry division was deployed in combination with other weapons and units, creating a greater need for exchange of information and coordination in order to ensure that the artillery began its barrage at the right moment, the tanks and the infantry met up at the right place and so on. This created huge demands for logistics which again increased the need for information and coordination. This created the larger 'tail' of the modern armed forces, but it also increased the complexity of operations as well as the military organisation as such. A post-First World War military is simply a much more complex organisation than a pre-First World War military. For the modern military to work, this organisation had to be managed at the same time as troops had to be led into combat.

This increased complexity is not unique to the US armed forces. In the German armed forces, the number of *Führungstruppen* (headquarter troops) increased five times between 1945 and 1975.[66] The German Army was previously famous for its focus on *Auftragstaktik*, which gave subordinate commanders the freedom to take the initiative and lead from the front. This involved a strong emphasis on leadership.[67] In Afghanistan, concludes Philipp Rotmann of the NATO Defence College, German troops displayed little ability to *Auftragstaktik* but had operations hampered by heavy management structures that required consultation with Berlin on many detailed matters that Moltke the Elder would have left to the commander on the ground.[68] It is hard not to see a connection between the increase in *Führungstruppen* and the decrease in *Auftragstaktik*.

In any system, decision-making consists of information exchange. The workings of the system are documented and described in terms of decision points to those with the authority to take decisions. In terms of decision-making styles, distinction can be drawn between leadership and management. As mentioned, leadership is the personal direction of operations by virtue of greater competence, knowledge and skill. Soldiers follow their sergeant because he impresses them with superior knowledge and experience. Thus, leadership is a personal quality exercised at personal risk, whereas management is a collective responsibility for making various processes work as smoothly as possible. Failure of a management process is a risk for the organisation rather than for the person in charge. This might be one of the reasons why, according to Thomas Ricks, American generals are less accountable for their actions today than during the Second World War. In *The Generals*, Ricks describes how the American generals became a managerial class rather than a team of leaders. Ricks' cardinal point is that American generals are

---

[66] Martin van Creveld, *Technology and War: From 2000 BC to the Present* (New York: The Free Press, 1991), 237.

[67] Stephen Bungay, *The Art of Action: How Leaders Close the Gaps Between Plans, Actions and Results* (London: Nicholas Brealey Publishing, 2011), 54–110.

[68] Philipp Rotmann, *Built on Shaky Ground: The Comprehensive Approach in Practice*, Research Paper, No 63, December 2010 (Rome: NATO Defence College, 2010), 7.

no longer relieved of their command when they fail to deliver the results expected.[69] This demonstrates the focus on running the military business rather than delivering a military effect. From this perspective, the military organisation communicates that the smooth running of the organisation is more important than delivering outcomes in the field.

The lieutenant leading his platoon receives and provides information differently from how staff officers managing a headquarters receive and provide information. In Sellin's description: 'For headquarters staff, war consists largely of the endless tinkering with PowerPoint scuds to conform with the idiosyncrasies of cognitively challenged generals in order to spoon-feed them information.'[70] Visiting the ISAF headquarters four years before Sellin was stationed there, British sociologist Anthony King also noted how 'schematics often in the form of graphics displayed by PowerPoint during meetings, played an important role in ISAF'.[71] Military organisations have fallen in love with PowerPoint; and perhaps that is no coincidence. Theodore Taptiklis describes how PowerPoint was Microsoft's DIY version of the slides carefully produced by McKinsey's art department in the 1970s before the advent of the personal computer made every man his own print shop. The McKinsey slides expressed a certain way of perceiving organisations and exchanging information with the management of these organisations. In Taptiklis' view:

> PowerPoint has mutated into a method and style of organization interaction that promotes formulaic thinking and mechanical expression and discourages participation and dialogue. Its very title now emphasizes the top-down, didactic orientation of its promoters.[72]

[69] Thomas E. Ricks, *The Generals: American Military Command from World War II to Today* (New York: Penguin, 2012).

[70] *Army Times*, 'The PowerPoint Rant that got a Colonel Fired'.

[71] Anthony King, *The Transformation of Europe's Armed Forces: From the Rhine to Afghanistan* (Cambridge University Press, 2011), 131.

[72] Theodore Taptiklis, *Unmanaging: Opening Up the Organization to Its Own Unspoken Knowledge* (London: Palgrave Macmillan, 2008), 55.

PowerPoints are about presenting facts in a few points, accompanied by tables and illustrations. As Colonel Sellin found out, the presentation is an end in itself. In Taptiklis' day, McKinsey realised that their real selling point was not the analysis of the strengths and weaknesses of an organisation, but the presentation of that analysis. This enabled McKinsey to standardise its analysis, thus creating huge economies of scale when the same points could be presented in different ways thus constituting different products. As the Commander's Update Assessment was presented endlessly in ISAF HQ, so was the need to streamline the organisation presented by McKinsey consultants to company after company. The presentation does not replace process, but as a means of organisational information the presentation places discussion, dialogue and innovation in the corridor outside the briefing room. 'A decision-maker sits through a 20-minute PowerPoint presentation followed by five minutes of discussion and then is expected to make a decision', T. X. Hammes notes, 'compounding the problem, often his staff will have received only a five-minute briefing from the action officer on the way to the presentation and thus will not be well-prepared to discuss the issues'.[73]

One should pronounce the slide program with an emphasis on the first word – it is about *power* rather than about *points*. The presenter might ask a question about a slide, but such questions remain rhetorical in the sense that it is very much the presenter's question rather than a question arising from any dialogue. PowerPoint presentations do not encourage dialogue or doubt, as Taptiklis points out. Quite the contrary, because the slides are presented, the information on the PowerPoints seems like facts independent of the person delivering the briefing. Where an oral presentation, a discussion group or even a memo have a clear and identifiable author, the points on the slides get their power from being truths in and of themselves. This independent nature of the PowerPoints is emphasised by how slide

---

[73] T. X. Hammes, 'Dumb-dumb Bullets', *Armed Forces Journal*, July 2009, www.armedforcesjournal.com/2009/07/4061641/ (10 May 2012).

templates ritualise information – in terms of which information fits the template and how that information is transmitted. King notes how the British officers under General Richards' command at ISAF HQ struggled with fitting the campaign plan into the schematic that was established on a notion of lines of operations that simply did not fit the complex Afghan situation. Tellingly, this presented a problem in itself for the staff officers; as Sellin described it, a staff officer's work was largely to produce the PowerPoints because, as King notes, 'without a simple collective representation for all the staff and for the other agencies outside the headquarters, it was very difficult to unify the disparate activities in Afghanistan'. Ultimately, the British staff officers produced a watered-down version of the plan that could be fitted into the PowerPoint templates.[74] The need to exchange information according to the templates and codes by which the system recognised itself thus became more important than actually communicating the Commander's intent.

Perhaps Colonel Sellin was just unfortunate to be assigned to the wrong headquarters. In 2005, when then-Colonel H. R. McMaster stabilised the city of Tal Afar in an operation that became the template for the US counterinsurgency strategy in Iraq, he banned PowerPoint presentations. Five years later, McMaster explained to the *New York Times* that a PowerPoint presentation is 'dangerous because it can create the illusion of understanding and the illusion of control – some problems in the world are not bullet-izable'.[75] In 2009–10, the renewed emphasis on a warrior ethos, which had come out of the Iraq War, led to a rebellion against the use of PowerPoint presentations in the US military; a rebellion that seemed to have few, if any, consequences for the number of slides produced. This demonstrates how the preference for PowerPoint presentations reflects the nature of information exchange in the modern military system.

---

[74] King, *The Transformation of Europe's Armed Forces*, 131–4.

[75] Elisabeth Bumiller, 'We Have Met the Enemy and He Is PowerPoint', *New York Times*, 26 April 2010, www.nytimes.com/2010/04/27/world/27powerpoint.html (18 October 2012).

PowerPoint presentations are a powerful symbol of how most military employees relate to war in the briefing room rather than on the front line. This is a product of the modern military system's focus on mobility and combined arms. This necessitated the development of capital-intensive forces as well as the 'back-office' functions to deploy and service these platforms. This is very expensive, but in the 1970s the United States made paying the bill a strategic objective in its own right that was to entrench America's dominant position in the defence sector. Norman Augustine describes how this has meant that the United States has invested more and more in fewer and fewer platforms. Perhaps the PowerPoint briefings are most of all a symbol of how finely tuned the US military has become in managing this particular form of capital-intensive warfare. However perfect that force may be, is it the right one at the right price? 'We must avoid being dominant and irrelevant at the same time', General James Mattis of the US Marine Corps notes, 'dominant in our chosen forms of war and irrelevant to the security of this country. So you've got to be able to adapt in the Darwinian sense.'[76] Needless to say, General Mattis made that presentation without using PowerPoints. The next two chapters will describe how the US military and its allies have to avoid irrelevance and extinction by adapting the force to new environment. There is little agreement on what constitutes the new environment, however.

[76] John Horgan, 'Cross-check: Obama's Choice for Warrior in Chief, Gen. James Mattis, Calls Iraq Invasion "the Dumbest Thing We Ever Did"', *Scientific American*, 27 July 2010, http://blogs.scientificamerican.com/cross-check/2010/07/27/obamas-choice-for-warrior-in-chief-gen-james-mattis-calls-iraq-invasion-the-dumbest-thing-we-ever-did/ (10 May 2012).

4   The counterinsurgency business model

On 25 March 2008, the Iraqi government launched an offensive in Basra in order to regain control of the city, which the British occupying forces in southern Iraq had largely left in the hands of the local Shia militia.[1] In Baghdad, the Americans had named the Shia neighbourhood of 2.4 million people after the Shia cleric Al-Sadr, and Al-Sadr was not about to let the attack in the south go unnoticed. He abruptly ended a ceasefire and his gunmen overran government checkpoints around Sadr City. At the same time, Shia fighters intensified their rocket and mortar attacks on the Green Zone, thereby directly challenging the authority of the government and coalition forces positioned there. The hard-won stability in Baghdad was in danger. In March 2007, 81 US service personnel had died in Baghdad, 39 in March 2008. In March 2007, 2,762 Iraqi civilians had died, 810 in March 2008.[2] These improved figures hardly rendered Baghdad a new tourist destination but did represent a serious success for the 'surge' of US troops into Baghdad in the spring of 2007. By creating a more secure environment, the US military hoped to provide the basis for reconciliation between the warring Iraqi factions, thereby creating the conditions for an eventual US withdrawal. On 25 March, Al-Sadr threatened to nullify the US successes towards creating stability and bring down the Iraqi government. The US Army had little choice but to intervene.

Colonel John Hort of the 3rd Brigade Combat Team, 4th Infantry Division, was entrusted with taking back Sadr City. The

---

[1] Thomas E. Ricks, *The Gamble: General Petraeus and the Untold Story of the American Surge in Iraq* (London: Penguin, 2010).

[2] iCasualties: Operation Iraqi Freedom and Operation Enduring Freedom Casualties, 2009, http://icasualties.org/ (26 September 2012).

army constructed a wall south of Sadr City to prevent Shia gunmen from filtering into the rest of Baghdad. As one US Army officer observed, Colonel Hort and his men did what any Roman legion would have done: they constructed fortifications around a besieged enemy, thus provoking him to storm their battlements instead of them having to storm his. However, the US Army had a few capabilities a Roman commander in Gaul could never have imagined. Hort's brigade was given direct control over two MQ-1 Predator drones, six AH-64 Apache helicopters, close air support from fighter jets and artillery support. Hort deployed his own men as snipers around the wall and sent out patrols in Stryker armoured combat vehicles.

In three months, Colonel Hort's brigade was able to defeat the Shia gunmen at the wall and gradually push into Sadr City. The brigade was able to do so because drones and other intelligence capabilities gave Colonel Hort a 'situational awareness', which his opponents lacked, even as they were fighting in their own back alleys. The firepower Colonel Hort had at his disposal to fight Al-Sadr's gunmen was perhaps comparable to a US Army division in the Second World War, which demonstrates how the operational unit for combat has become the relatively small brigade as compared to the larger division.[3]

The stakes in the battle for Sadr City could hardly have been higher for the US Army. The surge was the last chance to win the war in Iraq, but it was not only the fate of that war which hung in the balance. A new generation of American commanders had pledged their own futures and that of the Army on a strategy for winning the war in Iraq. They termed this strategy 'counterinsurgency', or, in one of the ever-present military acronyms, COIN. 'Today, when countering an insurgency growing from state collapse or failure', the COIN Field Manual stated, 'counterinsurgents often face a more daunting task: helping friendly forces reestablish political order and legitimacy

---

[3] David E. Johnson, M. Wade Markel and Brian Shannon, *The 2008 Battle of Sadr City*, prepared for the United States Army (Santa Monica, CA: RAND Corporation, 2011).

where these conditions may no longer exist.'[4] COIN was not only a strategy for winning the war on the streets of Baghdad – in the eyes of the 'coinistas', at least, it also provided an answer to the problems of the modern military system. A small but highly augmented unit fought the battle for Sadr City. This brigade combat team was not fighting the army of another state, but rather the gunmen of a religious leader with strong ties to a neighbouring country. While building the wall created mass and focus in the battle, the main objective was not merely defeating the enemy the way the Germans had been defeated on the western front, but instead conquering the 'hearts and minds' of the population. In the words of British General Sir Rupert Smith, it was a war 'amongst the people'.[5]

So much so, in fact, that daily life in Baghdad largely continued undisturbed while Hort's men were fighting Al-Sadr's gunmen. This is the nature of warfare in a world of globalisation, large-scale demographic shifts and urbanisation, the coinistas believe. The battle for Sadr City is widely regarded as the shape of things to come. If wars were previously fought on a battlefield, the conflicts of the future are to take place in a 'battlecity'. Conflicts are not being fought by large and expensive platforms, but instead by small units of soldiers, backed up by large intelligence operations and immense capabilities for indirect fire.

The COIN doctrine is one answer to how Western armed forces can become affordable for taxpayers and suitable for the conflicts of the future. As such, COIN constitutes a military business strategy, and this chapter will further explore the analysis upon which this strategy is based. At its heart, COIN focuses on the human resources of the military, and this chapter will thus focus on the role and identity of the soldier. While beginning with the internal debate within the US Army about COIN, it should nevertheless be noted that the

---

[4] US Army, *Counterinsurgency*, FM 3–24, MCWP 3–33.5, December 2006, www.fas.org/irp/doddir/army/fm3-24.pdf (7 June 2012), 1–21.

[5] Rupert Smith, *The Utility of Force: The Art of War in the Modern World* (London: Vintage, 2008).

fact that most of the United States' allies have been involved in the wars in Iraq and Afghanistan has rendered the COIN debate transnational. Twenty-three countries having lost service personnel in Iraq – from one soldier from Fiji to 179 from Britain – testifies to the involvement of other nations in how this and other wars are fought and organised. At various NATO seminars and national staff colleges, the pros and cons of COIN are being discussed. And in these settings, the stakes are as high as in the United States. Should the COIN argument prevail, the modern military system will be changed fundamentally. This is one reason why the COIN argument has been challenged in the post-Iraq and post-Afghanistan period. Without the immediate need to fight counterinsurgency campaigns, COIN seems abstract in theory and impractical for implementing the procurement priorities of the US and allied militaries. While the coinistas themselves may be on the defensive or in downright disgrace, as in the case of General Petraeus, their perspective on global developments is very much at the centre of the US Army and Marine Corps' argument for their future relevance. Their Strategic Landpower Task Force describes how the 'challenges of an uncertain future in an era of fiscal austerity requires intellectual rigor and honest debate'.[6] COIN provides one such rigorous and powerful narrative for the future of military power, because it reminds us that war is about people. As General Robert W. Cone, the head of the Strategic Landpower Task Force asserts, 'war is fundamentally a clash of human wills; technology is secondary'.[7] By focusing on the 'human domain', the coinistas point out the effect that demographic changes, technological connectivity and globalisation have on human interaction and identify warfare as one such interaction. This perspective connects the human resources of the planet with the human resources of the military.

[6] Strategic Landpower Task Force, *Strategic Landpower: Winning the Clash of Wills*, May 2013, www.ausa.org/news/2013/Documents/Strategic%20Landpower%20 White%20Paper%20May%202013.pdf (13 November 2013), 4.

[7] US Army, 'TRADOC: Strategic Landpower Concept to Change Doctrine', *US Army*, 16 January 2014, www.army.mil/article/118432/TRADOC__Strategic_ Landpower_concept_to_change_doctrine/ (28 January 2014).

The fact that the armies in the United States, Britain and continental Europe are reducing in size only confirms this point, because how to use the force that remains and justify the still substantial cost of that force becomes increasingly difficult within the framework of conventional battles. The austerity armies are too small to fight such old-fashioned wars and if they are to have a job other than as an 'army in being' they need to define their mission in terms of the future challenges of the 'human terrain'.

This chapter begins by laying out the coinista analysis of the challenges facing the modern military system as globalisation and large-scale demographic change. Secondly, the chapter turns to the answer offered by General David Petraeus and his fellow coinistas. This argument is found in the beginning of Petraeus's PhD dissertation from Princeton University, which lays out not only many of the COIN ideas but also how to implement them. This serves to show how the COIN argument is a restatement of a classical argument for using soldiers in other ways than those prescribed by the modern military system. This leads to an analysis of the very nature of the soldier as a human resource in the COIN doctrine and how that compares to business management notions of using human resources more creatively in civilian business. Lastly, I turn to the legacy of the COIN debate in order assess whether the coinistas have the blueprint for creating more agile and affordable armed forces.

PLANET OF SLUMS, PLANET OF ZOMBIES

Every form of warfare can refer to a constitutive battle – for example, Marathon (490 BC) for phalanx warfare or Midway (1942) for carrier warfare. Such constitutive stories are important, according to business strategist Richard Rumlet, because they constitute a kernel of the diagnosis that can 'replace the overwhelming complexity of reality with a simpler story, a story that calls attention to its crucial aspects'.[8] The story of a battle dramatises the choices, challenges and

---

[8] Richard Rumlet, *Good Strategy/Bad Strategy: The Difference and Why It Matters* (London: Profile Books, 2011), 81.

opportunities inherent in a certain form of warfare, and such stories define what Rumlet calls the 'domain of action'.[9] The story of the battle for Sadr City defines the urban domain of action in future conflict.

According to the UN, the global population reached 7 billion in 2012 and will increase to 9 billion by 2050. The additional 2.3 billion people will be born in the Global South, increasing its population from 5.6 billion in 2009 to 7.9 billion in 2050. This population increase will not only change the number of people on the planet, but also where they are living. In 2020, the UN estimates that most of the world's population will be residing in cities for the first time in human history. This will further contribute to the development of megacities – like Mexico City with 21 million inhabitants and Shanghai with 23 million. This increase in population may drive growth, but it will also contribute to the tendency for these large cities to embrace very different levels of development and income within the same urban space. Within these urban spaces, the unequal development of North and South will play out between the slums and the upscale neighbourhoods, which will often be in the form of gated communities. Mike Davis describes this as a 'planet of slums'.[10] Large elements of these cities will – alongside countries, of course – be part of what Paul Collier terms the 'bottom billon'. 'The countries at the bottom coexist with the twenty-first century', Collier explains, 'but their reality is the fourteenth century: civil war, plague, ignorance.'[11] Whatever the actual size of the 'bottom billion' in 2050 after the global population has increased by 2.3 billion people, the bottom billion is a condition of development that Collier sums up in the following: amongst the bottom billion, average life expectancy is 50 years, infant mortality rate 14 per cent, and 36 per cent of the children who make it beyond their first year of life will be malnourished. At a time when developing countries are boasting 4.5 per cent growth rates, the poorest countries are only

---

[9] *Ibid.*

[10] Mike Davis, *Planet of Slums* (London: Verso, 2007).

[11] Paul Collier, *The Bottom Billion* (Oxford University Press, 2008), 124–34.

growing by 0.5 per cent. Collier points out that these countries are actually poorer today than in the 1970s.[12]

From a military perspective, it is interesting to note that 73 per cent of those in the bottom billion live in a country marked by civil war or which has recently undergone civil war.[13] If these conditions are translated into megacities by the growth of population leading to urbanisation, then the UK Joint Concepts and Doctrine Centre concludes that 'rapid urbanisation is *likely* to lead to an increased probability of urban, rather than rural, insurgency. The worst affected cities *may* fail, with significant humanitarian and security implications.'[14] In the 1970s, revolutionary writers debated furiously whether the revolution was best staged in the city or the countryside.[15] Today, the issue is not so much where to stage a decisive battle for control over the nation. The Marxist perspective has been replaced by a more multifaceted perspective, which identifies a network of resistance against the central power for a number of reasons rather than a single movement inspired by a single ideology. A sprawling megacity offers rich possibilities for what Admiral James Stavridis terms 'convergent threats'. Admiral Stavridis has served has commander of the US Southern Command, which has had the war on drugs high on its operational agenda since the 1980s, and the European Command, which used to be focused on traditional, Cold War threats. Having held both jobs made the Admiral sensitive to how different types of security threats are 'converging'. The way in which organised crime and terrorists can join forces is 'really the dark end of the spectrum of globalization as you assess rising national security risks', Admiral Stavridis told the *New York Times*. 'I think that's a very possible and very dangerous business model, and you have to prevent narco-businessmen crossing those streams with the terrorists.'[16] An urban environment offers ample possibilities for convergence

---

[12] *Ibid.*   [13] *Ibid.*, 7–9.

[14] Ministry of Defence, *Global Strategic Trends: Out to 2040*, Strategic Trends Programme, 4th edition, 2010, www.mod.uk/NR/rdonlyres/6AAFA4FA-C1D3-4343-B46F-05EE80314382/0/GST4_v9_Feb10.pdf (28 August 2012), 12.

[15] G. J. Ashworth, *War and the City* (London: Routledge, 1991), 88–92.

[16] 'Globalizaton Creates a New Worry: Enemy Convergence', *New York Times*, 30 May 2013.

and provides access to the global infrastructure. This official awareness of the insurgent capabilities inherent in the new 'urban geography of fear'[17] puts a new urgency on the ability to control urban spaces. As G. J. Ashworth points out, there is a lot at stake. The city, especially a capital city, is a target-rich environment. 'The city presents a range of targets distinguished not only by their physical density but also by their practical and symbolic importance.'[18] What happens in the downtown of a capital city matters much more than what happens in remote rural areas. For that very reason, the level of conflict tolerated in a city is also much lower than in the countryside.

Just as the US Army invaded Sadr City in order to pacify it, the Brazilian Army moved into the Rio de Janeiro slums of Complexo do Alemão in the Christmas of 2010. The authorities referred to the 800-parachute infantry as a *Força de Pacificação* – a pacification force.[19] It is significant to note how the language the Brazilian government uses for a security operation against its own citizens in the country's largest city closely corresponds to the language of UN peacekeeping. Indeed, the Brazilian government can be said to follow Paul Collier's instructions for dealing with the plight of the 'bottom billion': military intervention.[20] According to Collier, the main problem in bottom billion societies is the lack of a stable societal framework. If local elites are corrupted, then foreign intervention and a neo-colonial administration from the outside might be the best and only option for getting the society back on track. In this light, it seems to matter little whether the local elites that fail to run their country properly are the bourgeoisie of Free Town or the gangs of the *favelas* of Rio.

This 'militarizing urban space'[21] is not isolated to Baghdad or Rio. In 2009, the armed forces were deployed against Mexico's drug

---

[17] Ashworth, *War and the City*, 91.

[18] *Ibid.*, 88.

[19] 'Occupation of Zona Norte Favelas', *The Rio Times*, 28 December 2010, http://riotimesonline.com/brazil-news/rio-politics/occupation-of-zona-norte-favelas/ (7 June 2012).

[20] Collier, *The Bottom Billion*, 124–34.

[21] Robert Warren, 'Situating the City and September 11th: Military Urban Doctrine, Pop-Up Armies and Spatial Chess', *International Journal of Urban and Regional Research*, 26(3) (September 2002), 614.

cartels. Ten thousand soldiers and federal agents patrolled Juarez on the US border.[22] In 2002, the US Army issued a field manual for urban combat emphasising this trend, arguing that 'the worldwide shift from a rural to an urban society and the requirement to transition from combat to stability and support operations and vice-versa have affected the US Army's doctrine'.[23] This transition from a combat role to supporting a civilian effort can be a long time coming. Just as the US military ultimately fought in Baghdad for seven years, the Brazilian paratroopers were still deployed to the *favelas* in 2012, waiting to be replaced by Police Pacification Units.[24] In 2014, the military presence in Rio de Janeiro's *favelas* was increased in order to pacify the neighbourhoods prior to the football World Cup.[25] The Brazilian red berets were patrolling the *favelas*, just like US soldiers were patrolling Sadr City, Mexican marines were patrolling in Juarez, and Israeli soldiers patrolling the West Bank. Add to this the image of French paratroopers patrolling the Kasbah in Algiers in the 1950s, and one gets the impression of how the situation appears to the coinistas. Sadr City is a constitutive battle because it epitomises the realisation on the part of the US Army that the central problem it will be asked to address is the instability of the Global South. In so doing, it redefines the history of the armed forces from a war-fighting organisation to a stabilisation instrument.[26]

To a Brazilian paratrooper, the conditions of the *favelas* obviously have a different urgency than the conditions of Sadr City have

---

[22] 'An Army Takeover Quells Violence in Mexico', *Washington Post*, 21 April 2009.

[23] M 3–06.11 (FM 90-10-1), *Combined Arms Operations in Urban Terrain*, Headquarters No. 3–06.11. Headquarters, Department of the Army, Washington, DC, 28 February 2002, 1–1.

[24] 'Military Forces Remain in Complexo do Alemão and Penha in Rio's Zona Norte (North Zone)', *The Rio Times*, 15 May 2012, http://riotimesonline.com/ brazil-news/rio-politics/military-forces-remain-in-complexo-do-alemao/ (7 June 2012).

[25] 'Brazilian Army Occupies Rio Shantytown Ahead of World Cup', CNN.com, 24 April 2014, http://edition.cnn.com/2014/04/24/sport/football/ brazil-world-cup-favela-slums/ (28 April 2014).

[26] Max Boot, *The Savage Wars of Peace: Small Wars and the Rise of American Power* (New York: Basic Books, 2003).

for an American soldier. However, the demographic trends add urgency to the problem of stabilising the rising megacities of the South. This urgency is an important part of the COIN storyline. The UN estimates that the population of the developed world will increase far less than in the South. In the developed countries in the North, the population is estimated to increase from 1.23 billion to 1.28 billion between 2009 and 2050. According to the UN estimates, this increase is in itself a reflection of the rise in the population of the southern hemisphere, as the UN projects a net migration from South to North of 2.4 million per year from 2009 to 2050. In other words, some 100 million people are expected to migrate north over the course of 40 years.[27]

These 100 million people are the zombies of the bestselling novel *World War Z*, which spent four weeks on the *New York Times* Bestseller List and was made into a movie starring Brad Pitt. This should in itself testify as to the popular appeal of a book insisting on treating the zombie theme as a national security issue. *World War Z* thus offers a perspective on how wide-ranging the kernel of the COIN narrative actually is, even though one obviously cannot regard *World War Z* as a publication on a par with the COIN Field Manual or conclude that the coinistas share the book's sentiments. The author, Max Brooks, has thus lectured on zombie preparedness for various US Army audiences, and the book is on a US Naval War College reading list.[28] The US Center for Disease Control (CDC) has a website dedicated to zombies.[29] Since the CDC features prominently for its failure to grasp the gravity of the zombie pandemic in most of the zombie literature, this website will bring a smile to the face of most zombie aficionados. But CDC's reason for engaging with the zombie phenomenon is serious enough. On the website, Assistant Surgeon General Ali S. Khan notes, 'if you are generally well equipped to deal with a zombie apocalypse you will be prepared for a hurricane,

---

[27] United Nations Population Fund, www.un.org/esa/population/publications/wpp2008/pressrelease.pdf (25 May 2012).

[28] Taffy Brodesser-Akner, 'Max Brooks Is Not Kidding About the Zombie Apocalypse', *New York Times*, 21 June 2013.

[29] Center for Disease Control, www.cdc.gov/phpr/zombies.htm (13 November 2013).

pandemic, earthquake, or terrorist attack'.[30] Thus, the zombie idea captures the anxieties of the age where the US Army and Marine Corps identifies 'the increased pace and mutability of human interactions across boundaries' as a source of insecurity.[31]

In *World War Z*, the zombie contagion spreads from China, where the authorities are unwilling to admit and unable to control the beasts mutating in the countryside.[32] This framing captures the Western fears of the rise of China in a distinct 'Yellow Peril' narrative updated to the twenty-first century. In many ways, *World War Z* is a globalisation narrative that tells the story from many different perspectives but also emphasises how the global infrastructure allows the zombies to spread. In the book, only the Israelis and South Africans have the good sense to defend themselves against the contagion in time. Societies less focused on their national security needs are brought to the verge of extinction. Again, a point is very clearly being made here.

A central theme of the novel is the inability of the US armed forces to deal with the zombie threat. The fact that this is an organisational inability becomes clear when the remainder of the US government and the surviving members of the public are regrouping in California. The government official responsible for distributing the very limited resources to feed refugees and develop a response to the zombies decides to take a hard look at the defence budget. He relates the contents of the discussion to the narrator in this way:

> 'You cannot mothball our Stealth bombers', they would yell. At first, I tried to reason with them: 'The M1 Abrams has a jet engine. Where are you going to find that kind of fuel? Why do you need a Stealth aircraft against an enemy that doesn't have radar?' … I could have gone head-to-head against the military for

---

[30] *Ibid.*

[31] Strategic Landpower Task Force, *Strategic Landpower*, 3.

[32] Max Brooks, *World War Z: An Oral History of the Zombie War* (New York: Three Rivers Press, 2006), 4–11.

the duration of the war, but I'm grateful, in the end, that I didn't have to. When Travis D'Ambrosia became chairman of the Joint Chiefs, he not only invented the resource-to-kill ratio, but developed a comprehensive strategy to employ it.[33]

In simple terms, the problem of the modern military system is the 'resource-to-kill ratio'. *World War Z* not only communicates the military problems, but also advances the coinistas' solution: organising the military with a focus on infantry and air support for ground units. In the book, it is able to reform itself into a fighting force capable of taking the battle to the zombies in a small town called – obviously – Hope, New Mexico. During the battle, soldiers listen to Iron Maiden as they shoot zombies in the head, which apparently is the only way to kill them – 'doctrine calls for one shot every full second'.[34]

World War Z captures Western fears in an era of large-scale demographic change and globalisation. The novel also relates the COIN narrative about how the armed forces are to respond to these challenges. This serves to demonstrate how the experiences of the Brazilian paratroopers in Rio and the US soldiers in Sadr City are connected to the dry UN statistics. *World War Z* is a cautionary tale, however, not unlike the 1950s science fiction of Robert Heinlein, who urged a strong commitment to defence against the Soviet Union.[35] The point about the zombies is that they are an abstract and absolute evil, and these abstractions must be translated into concrete action in order to deal with the real problems of the day; which is what the coinistas set out to do. In formulating their strategy, they focus on empowering the individual soldier, and in so doing, the coinistas formulate a very specific vision of the empowered and enlightened Western soldier as opposed to the zombie-like insurgents they are to defeat.

<hr>

[33] *Ibid.*, 144–5.   [34] *Ibid.*, 273–8.

[35] Ken Macleod, 'Politics and Science Fiction', in Edward James and Farah Mendlesohn, *The Cambridge Companion to Science Fiction* (Cambridge University Press, 2003), 232–3.

AN INSTITUTION TRANSFORMED BY WAR

On a cold and windy February day in 2011, the cadets of West Point received their commission. Secretary of Defence Robert Gates told them:

> When you receive your commission and walk off the parade field for the last time, you will join an army that, more than any part of America's military, is an institution transformed by war. The change has been wrenching for a service that a decade ago was essentially a garrison army, a smaller version of the Cold War force that faced down the Soviets in Europe and routed Saddam's divisions from Kuwait.[36]

For Secretary Gates, Sadr City and the surge had redefined the Army's mission. 'The army must confront the reality that the most plausible, high-end scenarios for the US military are primarily naval and air engagement',[37] the Secretary of Defense told the newly minted US army officers. If they wanted to pursue a career with heavy armour, the Secretary's words should make them pause. With the Army costing USD 141,728 million in 2011, one of the important factors rendering COIN attractive was that it was cheap. An active Army division costs around USD 5.2 billion, while an Army brigade combat team, like the one deployed in Sadr City, costs USD 1.2 billion.[38] The USD 4 billion difference is only saved, however, if the Department of Defense decides that it does not need the division in the future. Otherwise, the restructuring of forces to do counterinsurgency is, mostly, added to the bill. This was what happened during the wars in Iraq and Afghanistan.

---

[36] Robert M. Gates, Speech at United States Military Academy (West Point, NY). As Delivered by Secretary of Defense Robert M. Gates, West Point, NY, 25 February 2011. Defense.gov Speech, www.defense.gov/speeches/speech.aspx?speechid=1539 (24 January 2012), §5.

[37] *Ibid.*, §10.

[38] Michael O'Hanlon, *The Science of War: Defense Budgeting, Military Technology, Logistics, and Combat Outcomes* (Princeton University Press, 2009), 26.

In 2000, the US Army had a budget – in constant 2011 dollars – of USD 94,886 million. In 2010, the wars in Iraq and Afghanistan had increased that budget to USD 242,290 million (constant dollars). The aim of the Obama administration was to reduce that budget by USD 100,000 million by 2015.[39] In 2000, USD 150,681 million (constant) were spent on operations and maintenance by the entire Department of Defense. In 2010, that title had increased to USD 278,395 million (constant).[40] In a comparative perspective: from 2005 to 2010, the US operations and maintenance budget rose by 25 per cent, whereas the British budget for operations and maintenance increased by almost 16 per cent, and the French budget for this title rose by 10 per cent.[41] The difference illustrates the United States' much larger share of the operations, but the fact that the British and French operations and maintenance budgets increased significantly illustrates the added expense of conducting operations in places like Iraq and Afghanistan. If these kinds of operations are to shape the things to come, as the COIN establishment and Secretary Gates believe, then ministries of defence should expect to foot a larger bill.

This bill was by no means reduced by the fact that the operational focus shifted from the divisional to the brigade level. True, there was a lot of money to be saved by deploying smaller, less heavy units; however, it was equally true that inserting the soldiers into the precarious environments of Kandahar or Baghdad increased the need for personal capabilities and protection; a US soldier therefore wears gear worth USD 17,000.[42] Mine-resistant ambush-protected vehicles

---

[39] *National Defense Budget Estimates for FY 2011*, Office for the Undersecretary of Defense (comptroller), March 2011, http://comptroller.defense.gov/Portals/45/Documents/defbudget/fy2011/FY11_Green_Book.pdf (26 April 2012), Table 6-13.

[40] *Ibid.*, Table 6-11.

[41] *Ibid.*, Table 6-11; European Defence Agency, *National Breakdowns of European Defence Expenditure*, www.eda.europa.eu/Libraries/Documents/EDA_-_National_Defence_Expenditure_in_2005.sflb.ashx (25 September 2012), 20.

[42] David Barno, Nora Bensahel, Matthew Irvine and Travis Sharp, *Sustainable Pre-eminence: Reforming the US Military at a Time of Strategic Change* (Washington, DC: Center for a New American Century, 2012), 28.

(MRAP) cost USD 1.35 million each.[43] When the price of provisions, providing fire support and transporting an individual soldier was totalled, the US command in Afghanistan estimated the price of one soldier to be roughly USD 1 million.[44]

In order for the COIN strategy actually to provide a new business case for the armed forces, the COIN doctrine and capabilities would have to replace conventional doctrines and capabilities; not only in the Army, as suggested by the Secretary of Defense, but also in the Navy and Air Force. In these services, the 'resource-to-kill ratio' argument could also be applied. Why should the Navy float destroyers and aircraft carriers if the name of the game was to patrol unstable megacities? And why could these patrols not receive close air support from drones or traditional fighter aircraft? The insurgents in Sadr City did not have radar either, so why was the F-35 necessary?

Instead of focusing on platforms, the coinistas focused on people. 'The first and by far most important shift we must make is to stop emphasizing technology and start focusing on people', T. X. Hammes notes in one of the most influential books on COIN.[45] While the COIN narrative is focused on the changing nature of warfare, the domain in which the COIN commanders believe they must make a difference in order to transform the US military is human resources. In order to change organisations, Lee Bolman and Terrence Deal argue, organisations must be 'reframed': 'leaders have to find new ways to see things. They must also articulate and communicate their vision so others can learn to shift perspectives when needed.'[46] Bolman and Deal identify the human resource as one of reframing perspectives.[47]

<hr>

[43] Sharon K. Weiner, 'Organisational Interests versus Battlefield Needs: The U.S. Military and Mine-Resistant Ambush Protected Vehicles in Iraq', *Polity*, 42(4) (October 2010), 461–82.

[44] Todd Harrison, *Estimating Funding for Afghanistan*, 1 December 2009, The Center for Strategic and Budgetary Assessments.

[45] T. X. Hammes, *The Sling and the Stone: On War in the 21st Century* (Minneapolis, MN: Zenith, 2006), 232.

[46] Lee G. Bolman and Terrence E. Deal, *Reframing Organizations: Artistry, Choice and Leadership*, 3rd edition (San Francisco, CA: Jossey-Bass, 2003), 13.

[47] *Ibid.*, 111–32.

The human resource perspective constitutes a way to reframe the US Army – and by implication the entire US military. 'Today, America finds itself embroiled in 4GW in Iraq and Afghanistan as well as a worldwide struggle against al-Qaeda', Hammes notes, 'these are long-term struggles that will be won or lost primarily with human skills and knowledge'.[48] The counterinsurgency manual picks up on the human resource focus, arguing that

> Successful COIN operations require competence and judgment by Soldiers and Marines at all levels. Indeed, young leaders – so-called 'strategic corporals' – often make decisions at the tactical level that have strategic consequences. Senior leaders set the proper direction and climate with thorough training and clear guidance; then they trust their subordinates to do the right thing.[49]

This understanding of the soldier's role has massive implications for the organisation as such. The Army and Marine Corps should focus on developing the skill sets of soldiers and marines. 'As learning organizations', the COIN Field Manual notes, 'the Army and Marine Corps encourage Soldiers and Marines to pay attention to the rapidly changing situations that characterize COIN operations'.[50] A learning organisation is much more focused on its employees than on its technology. Defining themselves as learning organisations, the Army and Marine Corps try to address Hammes' complaint that official Department of Defense publications focus extensively on updating weapons systems but little on updating the personnel system.[51] To Hammes, the personnel system favours the officers who follow the path laid out by their superiors who are responsible for evaluating how the officer performs. You only get ahead, Hammes argues, by not rocking the boat; consequently, a successful officer is

[48] Hammes, *The Sling and the Stone*, 190.
[49] US Army, *Counterinsurgency*, 1–157.
[50] *Ibid.*, x.   [51] Hammes, *The Sling and the Stone*, 232.

thus most likely a risk-averse officer.[52] This creates organisations in which self-discipline stands in the way of learning and makes it very difficult to operate as a 'strategic corporal'. According to Hammes, the fact that officers rotate between postings means that they are in effect 'amateurs by profession'. Battalion commanders will often have fewer than six years in operational jobs when arriving at their command after some fifteen years of service.[53] Like Secretary Gates, Hammes wants officers to focus on fighting rather than various staff and administrative duties – he believes the officer corps could be halved using such measures.[54]

The COIN narrative is based on the notion that, at present, the defence establishment is focused on technology because it cannot escape the modern military system's way of thinking. Hammes notes that 'in the case of the United States, we have huge sunk costs in conventional forces'.[55] Hammes compares the Defense Department with IBM in the 1990s, when the company clung to old ways of doing business and was almost wiped out by the PC revolution. 'Like IBM, we are busy making things more like we think they should be, 3GW updated through Joint Vision 2020 and "transformation" rather than observing, understand, and reacting to the major changes in society.'[56]

For all its insistence on the transforming nature of warfare, the COIN argument is not new. At the time the modern military system found its form on the Western front, T. E. Lawrence was fighting a very different war in the Middle East: 'Ours seemed unlike the ritual of which Foch was priest', Lawrence wrote, 'in his modern war – absolute war he called it – two nations professing incompatible philosophies put them to the test of force',[57] Lawrence's Arab insurgency was about a very different philosophy of warfare: 'a province would be won when we had taught the civilians in it to die for our ideal of freedom', he writes.[58] In Lawrence's account, he has a fever

---

[52] *Ibid.*, 234.   [53] *Ibid.*, 235.
[54] *Ibid.*, 243.   [55] *Ibid.*, 201.   [56] *Ibid.*, 205.
[57] T. E. Lawrence, *Seven Pillars of Wisdom* (Ware: Wordsworth Editions, 1997), 179.
[58] *Ibid.*, 186.

in a tent and sweats his way through an epiphany on military strategy. As so much else in the *Seven Pillars of Wisdom*, it is an obvious literary ploy with clear reference to the prophet himself. This is precisely why Lawrence is such a great writer. He is able to convey the personal nature of his insight on warfare and tells the reader how this insight is so at odds with the conventional military doctrine that understanding the nature of insurgency warfare is a physical challenge to Lawrence.[59] This very personal approach is continued in much of the subsequent writing on COIN. The coinistas are those who have realised that things are changing, and they have little respect for the established procedures and hierarchies of the armed forces.[60] Paul Trinquier, the French COIN guru of the 1960s, is thus both ominous and scornful when he writes that 'if, like the knights of old, our army refused to employ all the weapons of modern warfare, it could no longer fulfil its mission'.[61]

From this perspective, the notion of generations of warfare can be read biographically as well as analytically. The notion of generations of warfare is to describe different types of strategy and technology. The first generation of modern warfare being Napoleon and the mass army, the second generation comes along with the rise of conscription, while the third generation refers to manoeuvre warfare and combined arms doctrine (in other words, what we refer to as the modern system). COIN represents fourth-generation warfare.[62] Analytically, the notion of generations provides a useful taxonomy of military history, but it is also useful for distinguishing between the old and new schools. It is a way to cast the proponents of heavy divisions and stealth bombers into knights polishing their armour, while at the same time presenting oneself as Swiss pike men or English archers, ready to do away with an outmoded way of warfare. Telling

---

59   *Ibid.*, 177–88.
60   Gian P. Gentile, 'A Strategy of Tactics: Population-centric COIN and the Army', *Parameters: The US Army War College Quarterly*, Autumn 2009, 10.
61   Roger Trinquier, *Modern Warfare: A French View of Counterinsurgency* (Westport, CT: Praeger, 2006), 90.
62   Hammes, *The Sling and the Stone*, 16–31.

the story in this manner also gives those who are able to forward the COIN argument a special status as the romantic outsider in the style of Lawrence and the soldier scholar in the style of Trinquier. When these roles are put into a human resource frame of reorganisation, you get a modern military hero. Meet General David Petraeus, who is the subject of the next section.

PETRAEUS AND THE 9/11 GENERATION

In September 2011, General David Petraeus was sworn in as Director of the CIA at a ceremony in the White House. 'You led the 9/11 generation', Vice President Joseph Biden told Petraeus, who was wearing a grey suit instead of his general's uniform, 'and turned it into the greatest group of warriors this country has ever seen',[63] The fourth-generation warriors had a clear leader in David Petraeus. This was no coincidence, since the political leadership was careful to identify Army officers able to conduct counter-insurgency and promote them to positions where they could influence how the Army conducted itself in the war.[64] Petraeus' new job as CIA Director was a way to ensure that the agency would continue to conduct its own shadowy war against al-Qaeda and other fourth-generation threats. Few could have been better prepared for the job than David Petraeus.

Petraeus wrote his PhD dissertation at Princeton not only on counterinsurgency, but on the US Army's organisational response to and rejection of counterinsurgency following the Vietnam War (1965–73). The dissertation is testament to the level of excellence the US military develops amongst its most able officers when it pays for their postgraduate education, but it is also a fascinating read, not only because it is well-argued and insightful postgraduate work, but also because Petraeus clearly used the opportunity to reflect upon the

---

[63] *New York Times*, 'US Defense Chief Tours 9/11 Memorial', 7 September 2011.

[64] David H. Ucko, *The New Counterinsurgency Era: Transforming the U.S. Military for Modern Wars* (Washington, DC: Georgetown University Press, 2009), 75.

nature of his organisation as well as the nature of warfare. This possibly explains why the dissertation remains unpublished.[65]

In the dissertation, Petraeus studies how the lessons of Vietnam were processed within the US military and how these lessons reflected military advice to the President in subsequent years. Petraeus is especially interested in how conventional wisdom is formed, wisdom which defines the organisation's response to a crisis, regardless of the particulars of the situation. Petraeus thus writes:

> The conventional wisdom, in other words, tends to be institutionalized. In the military, for example, the commonly accepted lessons of the past are institutionalised through personnel and resource allocations, through incorporation into doctrinal manuals, and through instruction in the military school system.[66]

Petraeus demonstrates how conventional wisdom in the US military excluded COIN from the toolbox it was able to present to the President. The US military did not want to plan for counterinsurgency: (i) because the generals only wanted to do so if a number of very specific and highly unlikely conditions were in place and (ii) because they remembered how operational requirements in Vietnam had eaten the budget for other material. 'The logical extension of this reasoning', Petraeus notes, 'is that forces designed specifically for counterinsurgencies should not be given high priority, since if there are not sizable forces suitable for counterinsurgencies it will be easier to avoid involvement in that type of conflict. An American president cannot commit what is not available.'[67] In so designing an organisational response to operations the US military really did not want to carry out, the military was further eroding the US ability to

---

[65] David Howell Petraeus, *The American Military and the Lessons of Vietnam: A Study of Military Influence and the Use of Force in the Post-Vietnam Era*, a dissertation presented to the Faculty of Princeton University in Candidacy for the Degree of Doctor of Philosophy, Woodrow Wilson School of Public and International Affairs, mimeo, October 1987, 20.

[66] *Ibid.*, 22.   [67] *Ibid.*, 305–6.

conduct COIN, Petraeus argues. The second-order effect of the lessons the military drew from Vietnam was to undermine close political–military integration, which is at the heart of waging a successful COIN operation. On the contrary, the US military was determined to avoid 'political "meddling" and "micromanagement" in operational matters'.[68] Petraeus dryly comments that while most officers can quote Clausewitz about the political nature of warfare, few seem to accept the implications of that quote completely.[69]

Petraeus thus knew why the US Army did not excel in counterinsurgency when he took command of the US Army Command and General Staff College at Fort Leavenworth, Kansas, in October 2005. Now he turned the analysis from his PhD dissertation on its head and created a new 'conventional wisdom' in the US Army. Conscious about how officers in the chain of command were socialised into a certain view on COIN, he engaged a network of reform-minded officers, a pensioned general and outside experts to tell the COIN story. Eventually that story was codified in the Army and Marine Corps Counterinsurgency Field Manual and Petraeus was tasked with carrying it out in 2007 – first in Iraq and then in Afghanistan.[70]

The human resource framing of the COIN doctrine not only places the soldier at the centre of attention, it focuses on the abilities of the human resource managers. Success does not depend on creating a new technology or product – it depends on developing the organisation's human talent. The ability to do so is a talent in and of itself. This human resource has been prevalent in business studies at least since the 1990s and contributed to creating a CEO celebrity culture,[71] and the public regarded Petraeus in much the same vein. The massive attention paid to his infidelity towards his wife in the autumn of 2012 seems to prove this. His love life was interesting only because

---

[68] *Ibid.*, 303.   [69] *Ibid.*, 303.

[70] Ricks, *The Gamble.*

[71] Stephen Bungay, *The Art of Action: How Leaders Close the Gaps Between Plans, Actions and Results* (London: Nicholas Brealey Publishing, 2011), 100–1; Theodore Taptiklis, *Unmanaging: Opening Up the Organization to Its Own Unspoken Knowledge* (London: Palgrave Macmillan, 2008), 41.

the public cared so much about his public persona, which was much undermined by the affair. As such, Petraeus is a military hero in a very different way from General Patton or even General Eisenhower, with whom he might otherwise share the most in common. Petraeus' success did not depend so much on winning battles or prevailing in campaigns as on transforming the organisation that fought the war. He was a war CEO. This was an approach to strategy much more recognisable to students of strategy at a business school than traditional strategic studies. Rumlet thus mentions Petraeus' Iraq strategy as an example of a strategy that successfully identified the critical factors and acted upon them.[72]

The successful COIN strategy shaped expectations. Petraeus was able to shape the enormous Army machine into a very precise instrument for a specific purpose. This makes it seems as though military success depends on the virtues and talents of a single leader capable of bending the world according to his will. Rumlet points out that this notion of leaders and leadership stems from a Protestant notion that 'one can impose one's visions and desires on the world'.[73] Petraeus himself seems to have a far more sophisticated notion of how the military system interacts with its environment, but that did not prevent the lionisation of him and his fellow coinistas. The notion of the general as the imposing leader stands very clear in Kaplan's portrait of General McChrystal, then commander in Afghanistan, in the *Atlantic Monthly*. The headline says it all: 'Man versus Afghanistan'.[74] Juan Manuel Santos, President of Colombia, echoes this sentiment writing about a very different counterinsurgency campaign: 'Colombia's experience in countering insurgency illustrates how much leadership and policy counts, and that history and culture

---

[72] Rumlet, *Good Strategy/Bad Strategy*, 3–4.
[73] *Ibid.*, 71–6.
[74] Robert D. Kaplan, 'Man versus Afghanistan', *The Atlantic*, April 2010, www.theatlantic.com/magazine/archive/2010/04/man-versus-afghanistan/307983/ (2 October 2012).

are not destiny.'[75] The next section further explores this will to power in the COIN concept.

THE FUNKY BUSINESS OF COUNTERINSURGENCY

In 2007, Colonel Burton, commander of Dagger Brigade Combat Team, 1st Infantry Division, in Iraq, described the officers in his command as 'our Nation's modern "300"'.[76] Colonel Burton's reference is clearly to the movie from 2006[77] rather than Herodotus' description of the Spartans' battle at Thermopylae. The comparison is significant because the movie places such emphasis on the prowess of the individual Spartan soldier. Far from being faceless members of a phalanx, they are strong individuals with perfectly toned, oiled torsos. This plays on the US Army's obsession with physical fitness, which is shared by its generals: Petraeus is renowned for being an avid runner;[78] General McChrystal is famous for only sleeping four hours a night, eating one meal a day and running seven miles each morning.[79] Whereas Herodotus' Spartan soldiers go to their destiny in accordance with Sparta's laws, the COIN doctrine appears to be more about realising individual potential and the power of the soldiers.

In Kaplan's 'Man versus Afghanistan' article, General McChrystal is presented as one of the 300, if not King Leonidas himself. In Kaplan's portrait, the General is a man capable of changing history by being a man: 'Stanley McChrystal's job is to serve as the deus ex machina for the rebirth of that modestly well-functioning mid-20th-century Afghan state.'[80] One man can change the fate of

[75] Juan Manuel Santos, 'Foreword', in David Richards and Greg Mills (eds.), *Victory Among People: Lessons from Counter Insurgency and Stabilising Fragile States* (London: RUSI, 2011), xiii.

[76] J. B. Burton, 'The Officer Critical Skills Retention Bonus', *Small Wars Journal*, 18 July 2007 http://smallwarsjournal.com/blog/the-officer-critical-skills-retention-bonus (24 January 2012).

[77] Warner Brothers, 2006.

[78] Ricks, *The Gamble*.

[79] Michael Hastings, 'The Runaway General', *Rolling Stone*, 2010, www.rolling-stone.com/politics/news/the-runaway-general-20100622?page=2 (23 May 2012).

[80] Kaplan, 'Man versus Afghanistan'.

an entire country by will alone. Obviously, Kaplan recognises that there are larger forces at play and that McChrystal is but one man in a massive military machine, but that only makes McChrystal and his mission more impressive. The complexities of Afghanistan and the Army are reduced to the choices of one man. This hyperbole captures the notion inherent in much of the management literature – that strong leadership is key to strategic agility and the faith that easily translates into the hero worship of the CEOs that are regarded as strong leaders.[81] A strong and creative leader can enable employees to become strong and creative. This is the basis of Jonas Ridderstråle and Kjell Nordström's *Funky Business*, a book about how 'talents make capital dance'.[82] In more mundane terms, their thesis is that, in a constantly changing global business environment, businesses depend on their human resources office more than their production line. Staying ahead of the competition is about agility and the ability to come up with new ideas before your competitor.[83] In other words, this is much the same analysis as presented by the coinistas. Where it is 'man versus Afghanistan' to Kaplan, to Ridderstråle and Nordström it is 'man versus market'.

Referring to American officers as the 300 is a way of striking what Christopher Coker describes as a 'Nietzschean pose'.[84] The Nietzschean focus on individual empowerment as suggested in much of the literature focusing on the human resource 'frame' receives maximum emphasis in Nordström and Ridderstråle's book. To them, succeeding in a modern business environment is all about striking a 'Nietzschean pose'. 'The world is a stage', they state. 'We all play roles – organisations as well as individuals. But instead of costume drama, we now have constant, unscripted, improvised theatre.'[85] Christopher

---

[81] Yves Doz and Mikko Kosonen, *Fast Strategy: How Strategic Agility Will Help You Stay Ahead of the Game* (Harlow: Wharton School Publishing, 2008), 123.

[82] Kjell A. Nordström and Jonas Ridderstråle, *Funky Business: Talent Makes Capital Dance* (London: Pearson, 2000).

[83] *Ibid.*, 22.

[84] Christopher Coker, *Barbarous Philosophers: Reflections on the Nature of War from Heraclitus to Heisenberg* (London: Hurst, 2010), 240.

[85] Nordström and Ridderstråle, *Funky Business*, 41.

Coker describes how Nietzsche decried the fact that he saw only soldiers, not warriors. Like Colonel Burton, Nietzsche wanted the modern 300 rather than the functionaries of the modern military system. This is reflected in the coinistas break with the notion of military personnel as mere parts of the modern military machine. The focus on the soldier's body – in the movies as well the fitness rooms of the US Army – is a way of emphasising Nietzsche's point that the warrior feels and connects with warfare in a direct, physical manner – something he can feel on his body – as opposed to the cold reality of the operator of a weapons system. 'Your enemy you shall seek, your war you shall wage – for your thoughts', Nietzsche writes.[86] War becomes a metaphor as well as something real, a need to strive beyond a mediocre day-to-day existence in favour of something greater. To Nietzsche, it is a question of being a warrior rather than factory worker.

Funky Business deals with the same point. Nordström and Ridderstråle argue that employees no longer work in order to fulfil an obligation, but rather to fulfil themselves.[87] This matters to Nietzsche, as well as to Ridderstråle and Nordström, because fighting for yourself – on the battlefield or in the boardroom – enables you to transcend your present condition. As Coker notes: 'Nietzsche's existential heroes are forever pushing back the boundaries of the world and finding themselves on the brink of an eternal truth, or are lost in contemplation of the emptiness at the heart of things.'[88] From a human resource perspective, this is the kind of employee you want if your agenda is change. Generals Petraeus and McChrystal need existential heroes to forge the US Army into a COIN force. Nordström and Ridderstråle term this 'directional leadership': 'Direction is not a matter of command and control, but of focusing, allowing and encouraging people to focus on what really matters.'[89] This is the heart of the COIN argument: the battle for Sadr City really matters as opposed

---

86 Coker, *Barbarous Philosophers*, 233.
87 Nordström and Ridderstråle, *Funky Business*, 183.
88 Coker, *Barbarous Philosophers*, 238.
89 Nordström and Ridderstråle, *Funky Business*, 187.

to the tank battles that more traditionally inclined officers in the US Army would like to focus on. The COIN agenda thus depended on individual officers and soldiers understanding their role in a new way; hence, the emphasis on the Army and Marine Corps as learning organisations. It was in the theatre that the Army and Marine Corps were to acquire new identities. This is reflected in Nordström and Ridderstråle's observation that 'the organisational paradigm means that the legal boundaries of the firm are of less and less importance. It is the operational boundary that counts.'[90] Furthermore, the coinistas would see this as confirmation of their belief that the globalised, urban threats of the future define an operational environment for the armed forces with the Army in the centre and the other services in support.

The Nietzschean reading of the human resource perspective presents COIN as something different from the technocratic version of the use of military power found in the COIN Field Manual. Colonel Gian P. Gentile commanded the 8–10 cavalry armoured reconnaissance squadron in Baghdad in 2006 and that experience did not make him respond well to the COIN Field Manual:

> I was angry and bewildered because the paradoxes, through their clever contradictions, removed a fundamental aspect of counterinsurgency warfare that I had experienced throughout my year as a tactical battalion commander in Iraq: fighting. And by removing the fundamental reality of fighting from counterinsurgency warfare, the manual removes the problem of maintaining initiative, morale and offensive spirit among combat soldiers who will operate in a place such as Iraq.[91]

Why eat soup with a knife when you can use a spoon? Colonel Gentile points out the unsaid in the COIN literature and the Field Manual: combat and killing. The Field Manual actually states that

---

[90] *Ibid.*, 184.

[91] Gian P. Gentile, 'Eating Soup with a Spoon', *Armed Forces Journal*, September 2007, www.armedforcesjournal.com/eating-soup-with-a-spoon/ (7 June 2012).

'warfare remains a violent clash of interests between organized groups characterized by the use of force'.[92] Combat remains a back-drop, which the real counterinsurgency effort plays out in front of, in most of literature. There are plenty of references to fighting in the literature as in the Field Manual. However, these references are usually followed by a 'but' that leads the reader's attention to the so-called 'non-kinetic' elements of the conflict. This leads Colonel Gentile and others to regard the Field Manual's version of reality as something less than war. The Nietzschean reading of COIN's human resource focus suggests that something else is going on. COIN allows soldiers to escape the confines of the modern military system and again become warriors. The coinistas recognise there is plenty of fighting in COIN operations, but it is not the fighting of conventional operations.

This means that the 'Nietzschean pose' is a spectacle in Guy Debord's meaning of the term. In *Society of the Spectacle*, Debord argued that, in modern society, 'all of life presents itself as an immense accumulation of spectacles. Everything that was directly lived has moved away into representation.'[93] In Debord's view, social relations are 'mediated by images'.[94] For example, automobile commercials define our idea of driving, and we are conscious of this as we drive on the motorway and find meaning in assuming the role of the guy driving a car in the commercial we saw the other night on television.[95] Debord's work is essentially about how the modern, capitalist society creates estrangement and how the void created by estrangement is filled with images of another, more fulfilling life – which can be bought at a store near you. This reflective mediation of the modern condition is also present in the COIN narrative. Debord would recognise COIN as an image of the politically savvy soldier who is able to operate in a civilian environment. This image sells the product of war to civilian experts and politicians, who appreciate

---

[92] US Army, *Counterinsurgency*, 1–1.

[93] Guy Debord, *Society of the Spectacle* (Detroit: Black and Red, 1983), 1.

[94] *Ibid.*, 4. [95] *Ibid.*, 37.

the sophisticated understanding of the role of military force and its use in the complexity of a globalised world. Another image entirely is used to sell COIN to the soldiers. This is the image of the 300. This dual framing works because the soldiers are to recognise the images as a mediation of their experience rather than the experience itself. The mediations also work because they actually supplement each other in as much as both images focus on choice and individuality. Thus, the individual soldier is able to regain the identity that is lost when they are merely regarded as one element in the heavy, modern military system. Instead of being one more grunt, the soldiers become warriors. Debord's deep insight is that since the soldier knows that they are wearing a mask – in effect buying the HR commodity, in Debord's terms – the warrior identity can be shaped to fit the moment. This seems false to Colonel Gentile, who insists on defining the essence of soldiering. For the coinistas, it offers the opportunity to strike two different poses at the same time without creating a contradictory narrative about the future of warfare. Special forces are at the centre of this narrative.

During the surge, when General McChrystal was head of special operations in Baghdad, the special forces operatives conducted ten raids per night in Baghdad, 300 a month. The General often took part himself. 'It was intense', he reminisced to Robert Kaplan in the *Atlantic Monthly*. 'We were hitting al-Qaeda in Iraq like Rocky Balboa hitting Apollo Creed in the gut.'[96] In McChrystal's mind, the war was not about two complex organisations slugging it out. Following the Nietzschean take on the human resource perspective, the fight could be reduced to a boxing match between individuals. This made the effort less abstract than it might seem when reading the COIN Field Manual. McChrystal had no doubt that there was real fighting going on, and al-Qaeda was every bit as scary as the Russian Ivan Drago of the Cold War. In *Rocky IV*, the complexities of the Cold War are reduced to two men settling an argument with

[96] Kaplan, 'Man versus Afghanistan'.

their fists.[97] McChrystal did the same thing when going out with his troops night after night in Baghdad. The conventional forces deployed by conventional tactics during the previous years in Iraq had been unable to close in on the enemy and settle the issue in a man-to-man conflict. For that is not what warfare is all about, according to the modern system. However, the COIN doctrine places the individual soldier and – not least – his commander at the centre of events. He is no longer merely a part of the military machine; he is a 'strategic corporal' or a general that gets his hands dirty. On the same note, engaging the enemy is no longer a large, alienating combined arms operation, but rather a series of specific attacks on specific houses in Baghdad – night after night. Even if McChrystal – at least rhetorically – likes to condense the combat in the streets of Baghdad to a fist-fight, he made the use of special operations forces on 'an industrial scale', his trademark in Iraq as well as Afghanistan. 'We are getting so good at various electronic means of identifying, tracking, locating members of the insurgency that we are able to employ this extraordinary machine, an almost industrial-scale counterterrorism killing machine', Lieutenant Colonel John Nagl has stated.[98]

The use of special forces has been a crucial element in most counterinsurgency campaigns. In Colombia, an important part of the fight against FARC was using special forces more offensively. In 2006, when President Uribe launched the fight, the military had not previously systematically targeted the FARC leadership. In the next five years, twelve of the eighteen persons in the leadership had been killed.[99] In Malaya, British special operation forces hunted down insurgents in a similar fashion. Likewise, it was paratroopers that raided the Kasbah in the Battle for Algiers. In Algeria and Indochina, the French realised that only the paratroopers could retain the initiative

---

[97] *Rocky IV*, MGM, 1985.

[98] 'Kill/Capture', *Frontline*, PBS, Transcript, www.pbs.org/wgbh/pages/frontline/afghanistan-pakistan/kill-capture/transcript/ (2 October 2012).

[99] Greg Mills, 'Coke Isn't It: Changing a Culture and Image of Violence', in David Richards and Greg Mills (eds.), *Victory Among People: Lessons from Counter Insurgency and Stabilising Fragile States* (London: RUSI, 2011), 398–9.

and discipline in the seemly unending struggle. Conventional units, which consisted mostly of conscripts, were left to provide mass, but commanders did not trust them with the strategically crucial jobs.[100] In Iraq and Afghanistan, the professional British and American armies have a different standard, but the fundamental issue on how to utilise conventional forces in an unconventional war remains. Perhaps that is why combat is only mentioned *sotto voce* in the COIN literature. The purpose of the COIN doctrine is to make the Army a 'funky business' by providing directional leadership. In order for that to work, every soldier and marine must feel as though they are one of the 300. In the 'unscripted, improvised theatre' that constitutes the COIN transformation of the Army, that role cannot be left to special operations forces alone.

Special forces are crucial for COIN operations and the narrative they constitute. This possibly means that special forces are not as special as they used to be, as reflected in the US budget for special operation forces. It was first in 1987 that the special operations forces became one of the defence programmes by which the Department measures its effort. At that time, USD 154 million were appropriated to special operations forces; in 2011, the number was USD 10,361 million. Conventional units are increasingly asked to perform tasks previously reserved for special forces. The US First Infantry Divisions are thus deploying troops to Africa to train forces in missions which would previously have been carried out by special forces.[101] Anthony King notes how special forces set the new standards within post-Cold War military organisations. If tank crews were the funky units of the 1980s that inspired the rest of the armed forces, then the Special Air Service (SAS) is the new black in the beginning of the twenty-first century. Interestingly, this has a transnational aspect, as the SAS and

---

[100] Martin Windrow, *The Last Valley* (London: Cassell, 2005); Alistair Horne, *A Savage War of Peace: Algeria 1954–1962* (New York: New York Review Books Classics, 2006).

[101] 'US Army Hones Antiterror Strategy for Africa, in Kansas', *New York Times*, 18 October 2013.

other elite European forces measure themselves against the US Navy SEALs or Delta Force.[102]

This corresponds to how Nordström and Ridderstråle describe how it is the creative departments rather than the production departments that set the standard for business operations. The focus on special forces is not only reflected in organisational culture, however; it is also reflected in how the military budgets are put together. King notes that, in relative terms, the airborne elements of the British armed forces have doubled in size as conventional forces have been cut. The Parachute Regiment, 5 Airborne Brigade and the Royal Marines have remained at the same strength as in the 1960s, while most other formations have been downsized.[103] One reason for this is that individual ability counts for more in a smaller force. 'With a limited budget, political will and manpower', Rune Henriksen argues, 'Western militaries are being forced to make sure every individual who is willing to fight is not just an able individual, but a very capable one.'[104] Nordström and Ridderstråle would probably find this reflects 'how talents are making the army dance'. This emphasises the human resource perspective at the heart of the COIN doctrine and how this redefines the role of the individual soldier.

HOOAH – THE COIN LEGACY

In Afghanistan in the 1880s, British Army personnel used the acronym HUA (heard, understood and acknowledged) when receiving orders from their commander. The US armed forces took up the term as a kind of verbal salute that can be shouted by a unit on the parade ground or by officers being briefed in an auditorium. In the Army, soldiers roar 'Hooah', as do Air Force personnel, whereas sailors bellow

---

[102] Anthony King, *The Transformation of Europe's Armed Forces: From the Rhine to Afghanistan* (Cambridge University Press, 2011), 200–1.

[103] *Ibid.*, 153.

[104] Rune Henriksen, 'Warriors in Combat: What Makes People Actively Fight in Combat', *Journal of Strategic Studies* 30(2) (2007), 187–223.

'Hooyah' and marines yell 'Oorah'. The counterinsurgency has not, as yet, received Hooah, Hooyah and Oorah all around.

The '9/11 generation' adopted counterinsurgency as a military strategy for winning in Iraq and Afghanistan as well as a business strategy for transforming the armed forces as such. The battles of Sadr City and Kandahar were the shape of things to come in a world defined by large-scale demographic change and globalisation, the argument went. 'I for one believe that our generation is in the midst of a paradigm shift, is facing its own "horse and tank" moment if you like', argued the Chief of the British General Staff General Sir David Richards.[105] Speaking in 2009 at the annual Land Warfare Conference, where British Army officers meet to discuss their profession, General Richards took a page from *World War Z*, arguing that 'how we deal with the growing disillusionment of 1.2 billion Muslims, most living in dangerously radicalised states, will be an issue that will dominate our professional lives'.[106] To generals like Richards, the battles of Sadr City or Helmand province ushered in a new generation of warfare, and the military system had to adapt. Specifically, this meant a focus on human resources as opposed to technology. Generals Petraeus, McChrystal and their colleagues in the US and in allied armed forces provided strong leadership for a generation that found that their training and equipment did not enable them to master the situation in the conflict the West became engaged in after 9/11. The conventional forces of the modern system were not dismantled; instead, they were organised in brigades rather than divisions, and the inspiration for these smaller units of organisation came more from special forces than combined arms. Special forces became the centrepiece of the COIN strategy. They provided the military muscle that enabled the stabilisation mission of the

---

[105] David Richards, 'Twenty-first Century Armed Forces: Agile, Useable, Relevant', 23 June 2009, RUSI, Whitehall, London, www.rusi.org/events/ref:E496B737B57852/info:public/infoID:E4A4253226F582/#.UoVSMRajPq8 (14 November 2013), §13.

[106] *Ibid.*, §15.

force as such. This created new confidence in the Army, but to the extent that the Air Force and Navy accepted the premise, the COIN concept provided clear guidelines for their operations as well.

The problem with the COIN concept is that it neither delivered on its promise of victory nor its promise concerning a more affordable defence establishment. When the US withdrew its forces from Iraq in 2012, John Nagl, who was a key player in writing the COIN Field Manual, concluded in the *New York Times* that perhaps 'messy and unsatisfying are the hallmarks of success in modern counterinsurgency wars'.[107] 'We're still dealing with the Taliban', Secretary of Defense Panetta told ABC News in May 2012; 'although they've been weakened, they are resilient'.[108] The problem was not so much that the situations in Baghdad and Kandahar were messy. The coinistas had never promised any more. On the contrary, the basis of their analysis was that the purpose of the armed forces was to provide a measure of stability in a messy world and help nudge the development in the right direction. As Colonel Messe of West Point, who was also involved in the work on the Field Manual, noted, 'warfare in a dangerous environment is ultimately a human endeavour, and engaging with the population is something that has to be done in order to try to influence their trajectory'.[109]

The problem for the COIN doctrine was that its main tenet had been a promise of empowerment. Soldiers and their commanders were to take control of their fate and remake the environment in which they fought as well as the Army in the course of the campaigns in Iraq and Afghanistan. That ambition largely failed. Colonel Gregory A. Daddis, also of West Point, noted, 'we're not really sure

---

[107] John Nagl, 'The Age of Unsatisfying Wars', *New York Times*, 6 June 2012, www.nytimes.com/2012/06/07/opinion/the-age-of-unsatisfying-wars.html?_r=1 (7 June 2012).

[108] 'This Week' Transcript: Defense Secretary Leon Panetta – *ABC News*, 2012: 2, http://abcnews.go.com/Politics/week-transcript-defense-secretary-leon-panetta/ story?id=16415968&page=2#.T9BMB451J0s (7 June 2012).

[109] Bumiller, Elisabeth, 'West Point is Divided on a War Doctrine's Fate', *New York Times*, 27 May 2012, www.nytimes.com/2012/05/28/world/ at-west-point-asking-if-a-war-doctrine-was-worth-it.html (7 June 2012).

right now what the Army is for'.[110] As defence cuts became the norm in NATO after 2008, the 'COIN argument' was no longer sufficient to merit continued funding. 'The likelihood of further military expeditions is diminishing', John Mackinlay observes, 'and their enormous cost in treasure and manpower sits badly with their tangible but minimal benefit.'[111] All the more so, because Britain and the United States invested in expeditionary COIN operations without disinvesting previously approved conventional procurement programmes. As David Ucko and Robert Egnell point out with reference to Britain, 'it illustrates that while the MoD [Ministry of Defence] spent its intellectual energy and finite budget on high-tech systems that it had trouble fielding or finding a use for in the modern battlefield, it had underinvested in the basic gear needed for wars as they presented themselves'.[112] The COIN focus on basic infantry skills did not sit well with the modern system approach, which meant that when the operational need for COIN capability receded, COIN capabilities were not seen as a way to save money but instead became yet another area where money should be saved. The British Army was cut by 20 per cent and reduced to a size that was not encouraging if one was bent on carrying out long-lasting COIN operations. The British Army thus states that fulfilling its share of the budget cuts 'required a significant change to the current structure of the Army which has most recently been optimized for enduring operations in Afghanistan'.[113] That is not to say that the British Army does not expect to be engaged in stability operations, but the purpose of the reform is to ensure that it can do other operations than COIN. In the British case, the operations in Iraq and Afghanistan are not regarded as drivers of change, as

---

[110] *Ibid.*

[111] John Mackinlay, 'After 2015: The Next Security Era for Britain', *Prism* 3 (2012), 53.

[112] David H. Ucko and Robert Egnell, *Counterinsurgency in Crisis: Britain and the Challenges of Modern Warfare* (New York: Columbia University Press, 2013), 142.

[113] British Army, *Modernising to Face an Unpredictable Future: Transforming the British Army*, July 2012, www.army.mod.uk/documents/general/Army2020_brochure.pdf (27 September 2012).

in the COIN narrative, but rather one type of operation among many. The Russian intervention in Ukraine in 2014 underlined that point to military planners. This means that the British Army cannot save money by focusing on COIN, but must instead make a business case on being more 'agile'. In practice, that means a much smaller force on a high readiness and greater reliance on reserves.

COIN failed to deliver on the most important parameter: reducing defence costs. As General Karl Eikenberry, who served as ambassador in Kabul, notes in his autopsy of the Afghanistan war, 'unbounded and unconstrained by civilian authorities, the commanders of the incredibly well-funded US armed forces fixated on a way forward that was breathtakingly expansive and expensive'.[114] The coinistas had identified the shortcomings of the heavy conventional system and pointed to new, more agile formations with special forces as the tip of the proverbial spear. However, these new forces and capabilities were largely placed on top of existing structures; structures that became even more expensive, as the 'strategic corporal' had to be equipped for operations 'amongst the people' and sustained for 'long wars'. Instead of concentrating the US defence budget, the coinistas provided the rationale for the greatest expansion of the budget since the Korean War.

[114] Karl Eikenberry, 'The Limits of Counterinsurgency Doctrine in Afghanistan', *Foreign Affairs*, September/October 2013.

5 Converging technologies business model

For months, miniature sensors had been swarming Iran. Guards at the surface-to-air missile (SAM) sites and radar installations had been absent-mindedly slapping at insects that were actually miniature, biologically engineered observation platforms. The platforms had swapped data with each other and the operatives back in the United States who were hacking the CCTV cameras and other civilian information networks. These enormous quantities of information were processed and distributed to the unmanned combat air vehicles (UCAV) when they entered Iranian territory. The drones were able to approach at hypersonic speed, velocities that a human pilot would never be able to endure, identifying and firing on their targets with an accuracy a human pilot would rarely be able to match. Some of these platforms were able to fire directed-energy weapons, while others fired multiple types of smart munitions and yet others were smart munitions themselves. The UCAVs also fired high-powered microwaves that disabled parts of the Iranian air defences and communication networks. The smallest platforms were created using nanotechnology; and thousands of them were able to enter and destroy the underground nuclear facilities by entering through ventilation shafts or cracks in the concrete, while larger units combined in large attacks on SAM sites when needed and dispersed to avoid counterattacks from the Iranians. The machines did not fight on their own, though; automated Ospreys inserted a special forces unit close to the nuclear sites at Esfahan in the first hours of the operation. These soldiers had their biochemistry modified to keep them alert at all times during the stressful operation; not that they would feel stressed, since their biochemistry had also been modified to withstand stress and better endure physical injuries.

None of these capabilities made the battle of Esfahan a transformational battle in the way of the Battle of Cambrai of 1917, when tanks were first used in battle in significant numbers. These technologies had been foreseen in the National Science Foundation report about the convergence of nano, biological, information and cognitive (NBIC) technologies in 2003. What rendered the battle transformational was the fact that the US forces came under attack when the Iranians introduced the NBIC bomb. The soldiers were hit first by the bombs, which were basically a nano-swarm delivered by an old-fashioned RPG. The nanoparticles hacked their way into the soldiers' information- and communication equipment, while other particles attempted to turn their biochemical augmentation into malignant viruses that attacked the American soldiers' immune systems. One of the soldiers had failed to upgrade his security protocols and the Iranian malware was transmitted from his equipment to the entire US system. From that point on, the centre of gravity of the battle was not the kinetic skirmishes on the ground and in the air but rather the race between the American and Iranian systems engineers. The cyberwarfare quickly came to include attacks on civilian networks in Iran and the West. When Walt Whitman coined the term 'wholesale engines of war', he could hardly have known how well it would fit the NBIC technologies.[1]

Imaginary battles have been a central element in Western discussions about the future of warfare since *The Battle of Dorking* was published in the summer of 1871. In this imaginary battle, the Kaiser's army destroyed an ill-prepared and inadequately equipped British Army. In conjuring up this national calamity, Lieutenant Colonel Sir George Tomkyns Chesney made the case for a larger and better-equipped Army to deal with capabilities the Germans had just demonstrated in the invasion of France in 1870.[2] Sir Chesney made a lot of money from the publication of the story of this imaginary

[1] Quoted in I. F. Clark, *Voices Prophesying War 1763–1984* (Oxford University Press, 1966), 70.
[2] *Ibid.*, 30–48.

invasion in *Blackwood's Magazine*. This added motivation to other military experts or concerned patriots who wanted to draw attention to the need for military innovation. 'It became standard practice for writers in the major European countries to describe the shape of the war-to-come in order to demonstrate the need for bigger armies or better warships', I. F. Clark explains in his magisterial *Voices Prophesying War*.[3] Imaginary war became a genre that enabled the public and experts alike to discuss the military consequences of technological development. Chesney was an engineer and therefore belonged to a part of the British Army that was much more focused on the new possibilities of science than on the merits of the Army's past. He thus suggested his tale of a German invasion to publisher John Blackwood as 'a useful way of bringing home to the country the necessity for a thorough reorganisation'.[4] Chesney represented the prototype of a new generation of officers cradled in technological innovation. 'War is being drawn into the field of the exact sciences', H. G. Wells noted.[5]

'[W]ar itself became an exercise in managing the future', so noted military historian Martin van Creveld, and he observes further that 'the most successful commanders were not those most experienced in the ways of the past but, on the contrary, those who realized that the past would not be repeated'.[6] From this perspective, it makes a difference that people inside and outside the military are making NBIC technologies part of imaginary wars, and it is these imaginings I tried to evoke in the introduction to this chapter. The technology described in the imaginary version of a US intervention in Iran is based on the factors identified by the National Security Panel in the National Science Foundation report from 2003 on NBIC technologies. The Panel mentions seven factors: (i) data linkage, threat anticipation and readiness, (ii) uninhabited combat vehicles, (iii) warfighter education and training, (iv) chemical/biological/

---

[3] *Ibid.*, 3.   [4] *Ibid.*, 47.   [5] Quoted in *Ibid.*, 100.

[6] Martin van Creveld, *Technology and War: From 2000 BC to the Present* (New York: The Free Press, 1991), 218.

radiological/explosive (CBRE) detection and protection, (v) warfighter systems, (vi) non-drug treatments for enhancing human performance and (vii) applications of brain–machine interface.[7] With regards to the latter, one should imagine the Iranian and American systems engineers being wired into the information systems they are to operate, thus rendering them able to react to malware intuitively rather than by typing commands.

To many of those in the Western armed forces, these are neither dreary bullet points in a National Science Foundation report nor science fiction; but rather, they are something almost attainable as a result of the possibilities opened up by the convergence of NBIC technologies for transforming the military system. In 2010–12, drones arrived as a serious item on the public agenda. The Ngram Viewer on Google Books shows that references to unmanned aerial vehicles (UAV) took off after 2002 – at the same time the US military began deploying them extensively in Iraq and Afghanistan. References to the armed drones (UCAV) seem less prevalent.[8] In 2009, Peter Singer published *Wired for War*, which introduced the 'robotics revolution' to a non-specialist audience.[9] The trade journal *Jane's International Defence Review* launched a section on unmanned systems in the autumn of 2010.[10] While the public discussed drones, the national security establishment was discussing cybersecurity. 'Cyber attacks are becoming more frequent', NATO's Strategy Concept from 2010 asserts, 'they can reach a threshold that threatens national and Euro-Atlantic prosperity, security and stability. Foreign militaries and intelligence services, organised criminals, terrorist and/or extremist

[7] Mihail C. Roco and William Sims Bainbridge (eds.), *Converging Technologies for Improving Human Performance: Nanotechnology, Biotechnology, Information Technology and Cognitive Science*, National Science Foundation report (Dordrecht: Kluwer Academic, 2003), www.wtec.org/ConvergingTechnologies/ Report/NBIC_report.pdf (27 September 2012), 327–30.

[8] Google Ngram Viewer. 2012. Google Ngram Viewer, http://tiny.cc/laoclw (28 September 2012).

[9] Peter W. Singer, *Wired for War: The Robotics Revolution and Conflict in the 21st Century* (London: Penguin, 2009).

[10] *Jane's International Defence Review*, 43 (October 2010).

groups can each be the source of such attacks'.[11] In October 2012, Secretary of Defense Leon Panetta stated that US commercial servers had been targeted by distributed service-denial attacks. On the surface, these attacks might appear quite harmless. Not so to Secretary Panetta, however, who imagined these attacks as a foretaste of future wars: 'These attacks mark a significant escalation of the cyber threat and they have renewed concerns about still more destructive scenarios that could unfold. For example, we know that foreign cyber actors are probing America's critical infrastructure networks.'[12] Making that argument, the Secretary of Defense wanted to put cyberwarfare high on the security agenda. Despite the focus on cybersecurity and drones, few outside specialist circles seem to realise that drones are merely a precursor of technologies that may result in the autonomisation of warfare and that this autonomisation is closely linked to the information technologies upon which cyberwarfare capabilities depend. In other words, the technologies defining future warfare converge with a number of other technologies, thus creating a technological transformation that is of a more profound nature than the individual technological innovations suggest at first glance. It is a matter of temperament whether one believes these technologies will merely define new capabilities or whether the new technologies will constitute a 'singularity' in the sense of creating qualitatively different conditions for human existence – on the battlefield as well as in civilian society.[13]

---

[11] *Strategic Concept For the Defence and Security of The Members of the North Atlantic Treaty Organisation*, adopted by Heads of State and Government in Lisbon, November 2010, www.nato.int/lisbon2010/strategic-concept-2010-eng.pdf (3 October 2012), §12.

[12] Leon Panetta, *Defending the Nation from Cyber Attack (Business Executives for National Security)*, as Delivered by Secretary of Defense Leon E. Panetta, New York, 11 October 2012, www.defense.gov/speeches/speech.aspx?speechid=1728 (30 October 2012), §20.

[13] Vernor Vinge, *The Coming Technological Singularity: How To Survive in the Post-Human Era*, 1993, http://mindstalk.net/vinge/vinge-sing.html (13 September 2012).

Whatever the scope of the NBIC transformation, it offers opportunity for Western armed forces to escape Augustine's law. Procurement budgets are rising and the number of platforms falling because existing technologies cannot support the military's need for combined arms platforms within the budget that Western nations in general and the United States in particular are able to allocate to their armed forces. NBIC technologies can reverse this trend by changing the technological side of the equation. With convergence technologies, the armed forces will, for the first time in eighty years, be able to procure more platforms for the same, or perhaps an even smaller, budget. It all depends on the ability of the armed forces to internalise the convergent technologies into a reform of the military business model. If the military system remains in its modern form, the net result might only be that, in 2040, the USA is able to field a single fighter aircraft and one UCAV instead of the one fighter aircraft that Augustine predicted. The crux of the matter is that NBIC technologies can only break Augustine's laws by making autonomous platforms central to how the military goes about its business.

Convergence technologies are not there yet, however. The first order of business is therefore examination of how contemporary military strategists are using research and development as a way of innovating the armed forces. Firstly, this chapter deals with the American AirSea Battle concept, which challenges the COIN doctrine's notion of how the security environment is developing. This concept focuses on procuring the technologies that – by themselves and by the force of the US defence budget – secure the United States and its allies in the geopolitics of a globalising world. This leads to a focus on procurement, which is the second issue dealt with in the chapter. The military is not attracted to convergence technologies only because they may solve the budget crisis, however; convergence technologies fit a technocratic approach to warfare that is very much a part of the focus on technology and procurement that became entrenched as part of Marshall's strategy for outspending the Soviets in the 1970s. This is the subject of the chapter's third section. The concluding section deals with the question of how the armed forces are to manage the transformation from present to future capabilities.

AIRSEA BATTLE AND THE CONVERGENCE

In the United States, the AirSea Battle concept represented the Air Force and Navy's answer to the decade of counterinsurgency. In terms of the politics of the Department of Defense, the implication of the name was clear: just as the concept of AirLand Battle had reconfigured the US armed forces after the Vietnam War, so would AirSea Battle change the focus from counterinsurgency to the 'real' strategic challenge in the Pacific.[14] AirSea Battle defined the core business of the armed forces in terms of state-to-state confrontation as opposed to COIN's state–non-state-actor blueprint for future conflict. Yet both concepts were based on in-depth analysis of the trends that will probably shape the future conflict environment.[15] Just as COIN is not merely a recipe for defeating the insurgencies in Iraq and Afghanistan, AirSea Battle is more than a strategy for containing and, if necessary, defeating the rising Chinese power. COIN is a business case for the military business in the twenty-first century based on the premise of large-scale social change (demography and globalisation). Similarly, AirSea Battle is a longer-term business model because its premise is that current technological developments will accelerate and, in time, create completely new conditions for war. These technological developments are what prompted the need for the AirSea Battle concept in the first place, according to one official presentation of the background of the concept:

> Over the past two decades, the development and proliferation of advanced weapons, targeting perceived U.S. vulnerabilities, have the potential to create an A2/AD environment that increasingly

---

[14] Jan van Tol, Mark Gunzinger, Andrew F. Krepinevich and Jim Thomas, *AirSea Battle: A Point-of-Departure Operational Concept* (Washington, DC: Center for Strategic and Budgetary Assessment, 2010), 5–8.

[15] *Ibid.*

challenges U.S. military access to and freedom of action within potentially contested areas. These advanced systems encompass diverse capabilities that include ballistic and cruise missiles; sophisticated integrated air defense systems; anti-ship weapons ranging from high-tech missiles and submarines to low-tech mines and swarming boats; guided rockets, missiles, and artillery, an increasing number of 4th generation fighters; low-observable manned and unmanned combat aircraft; as well as space and cyber warfare capabilities specifically designed to disrupt U.S. communications and intelligence systems. In combination, these advanced technologies have the potential to diminish the advantages the U.S. military enjoys in the air, maritime, land, space, and cyberspace domains today. If these advances continue and are not addressed effectively, U.S. forces could soon face increasing risk in deploying to and operating within previously secure forward areas – and over time in rear areas and sanctuaries – ultimately affecting our ability to respond effectively to coercion and crises that directly threaten the strategic interests of the U.S., our allies, and partners.[16]

This is more a shopping list than a strategy; a description of what the Chinese have bought and what the United States will need to buy to keep up.[17] The length of the list is a point in and of itself. Itemised like in the quote above, one is presented with a blur of new platforms and capabilities that challenge the current US technological superiority and thus threaten to shift the balance of power in the western Pacific. If the ability to conduct combined arms operations is central to the modern military system, then the new Chinese capabilities are truly disabling when the items on the list are tallied. The US response cannot merely be to make its own shopping list, but

<hr>

[16] Marines.mil – the Official Homepage of the United States Marine Corps, www.hqmc.marines.mil/News/NewsArticleDisplay/tabid/3488/Article/78942/the-air-sea-battle-concept-summary.aspx.

[17] Barry D. Watts, *The Maturing Revolution in Military Affairs* (Washington, DC: Center for Strategic and Budgetary Assessments, 2011), 15–25.

instead to look at these new capabilities and the countermeasures in total as a 'technology-enabled 21st century operational environment'.[18] When adding up the items on the list in this manner, one is presented with a technological challenge to how warfare is carried out today. Perhaps this perspective is the crucial difference between AirSea Battle and COIN. Where the coinistas wanted to reframe the armed forces by means of human resources, the 'technoistas' focus on the system.

The systemic approach to military transformation follows from how the technologies listed in the description of AirSea Battle function together. For someone schooled in the modern military system, this line of thought comes naturally. Thus, the proponent of technological transformation in the 1990s argued that information technologies created a 'system of systems' that reduced friction, thus making command and control easier. This constituted a 'revolution in military affairs' that would enable the United States to realise the full potential of the modern military system.[19] The National Science Foundation also refers to a revolution in its 2003 report that introduces the notion that the impact of developments in nanotechnology, biotechnology, information technology and new technologies of cognitive science should be seen as one, rather than as separate developments. Taken together as NBIC, these developments could transform society profoundly, the report argues:

> In the early decades of the 21st century, concentrated efforts
> can unify science based on the unity of nature, thereby
> advancing the combination of nanotechnology, biotechnology,
> information technology, and new technologies based in cognitive
> science. With proper attention to ethical issues and societal
> needs, converging technologies could achieve a tremendous

---

[18] US Department of Defense, *Quadrennial Defense Review 2014*, March 2014, www.defense.gov/pubs/2014_Quadrennial_Defense_Review.pdf (28 March, 2014) 6.

[19] Bill Owens (with Ed Offley), *Lifting the Fog of War* (New York: Farrar, Straus and Giroux, 2000).

improvement in human abilities, societal outcomes, the nation's productivity, and the quality of life. This is a broad, cross-cutting, emerging and timely opportunity of interest to individuals, society and humanity in the long term.[20]

The authors of the report do not merely speak of the developments in the laboratory. These innovations have already moved from the lab into the market. The global market for nanotech products is expected to increase tenfold from 2001 to 2015. This is reflected in an increase in worldwide public investment in nano research and development from USD 500 million in 1997 to USD 4.5 billion in 2004. In 2004, private research and development in nanotech in Europe, the United States and Japan accounted for 3.7 billion dollars.[21] In the United States, biotech revenues in 2000 were USD 23 billion; in 2005, this had more than doubled to 50 billion.[22] The market for artificial intelligence (AI) was USD 21 billion in 2007 and growing by 12 per cent annually.[23] The US military was funding 81 per cent of American AI research. Explanation is found in the conclusions of the National Security Panel of the National Science Foundation report:

> Investment in convergent nanotechnology, biotechnology, information technology, and cognitive science is expected to result in innovative technologies that revolutionize many domains of conflict and peacekeeping. We are entering an era of network-centric combat and information warfare. Increasingly, combat vehicles will be uninhabited, and robots or other

[20] Roco and Bainbridge, *Converging Technologies for Improving Human Performance*, ix.

[21] Frank Simonis and Steven Schilthuizen, *Nanotechnology: Innovation Opportunities for Tomorrow's Defence*, 2006, www.futuretechnologycenter.eu/downloads/nanobook.pdf (6 November 2012), 8–10.

[22] Ministry of Defence, *Global Strategic Trends: Out to 2040*, Strategic Trends Programme, 4th edition, 2010, www.mod.uk/NR/rdonlyres/6AAFA4FA-C1D3-4343-B46F-05EE80314382/0/GST4_v9_Feb10.pdf (28 August 2012), 143.

[23] Singer, *Wired for War*, 78.

automated systems will take on some of the most hazardous missions.[24]

The National Science Foundation carefully states the case for the NBIC revolution in positive terms as an opportunity; but the report leaves the reader with a clear sense that there is a bus to miss if the possibilities of NBIC convergence are not utilised by a convergence of national effort. In the view of the National Security Panel, the basic analysis of the coinistas is correct. 'The United States and its closest allies represent only a small fraction of the world population', the Panel asserts, 'and in the asymmetrical conflicts of the early 21st century, even a small number of dedicated enemies can cause tremendous damage',[25] If demographics and globalisation are going to make the West a small and, relatively speaking, poorer part of the globe, then the West must use technological innovation to compensate for this development. This way of thinking is echoed in the analysis behind AirSea Battle. NBIC represents a threat as well as an opportunity, and if the United States and its allies are unable to utilise the opportunity, then their military capabilities will be as relevant as the Polish cavalry against German tanks in 1939. In 2010, the US *Quadrennial Defense Review*, which sets the Pentagon's priorities every fourth year, clearly identified the need to develop future capabilities and to do so in a manner reflecting actual needs and the capacity of the US taxpayer to foot the bill.

> [The] Quadrennial Defense Review advances two clear objectives. First, to further rebalance the capabilities of America's Armed Forces to prevail in today's wars, while building the capabilities needed to deal with future threats. Second, to further reform the Department's institutions and processes to better support the urgent needs of the warfighter; buy weapons that are usable,

---

[24] Roco and Bainbridge, *Converging Technologies for Improving Human Performance*, 327.

[25] *Ibid.*, 329.

affordable, and truly needed; and ensure that taxpayer dollars are spent wisely and responsibly.[26]

These two goals obviously do not support each other. The not-too-subtle point made by the proponents of AirSea Battle is that while the Department of Defense under Secretaries Rumsfeld and Gates focused on the 'urgent needs of the warfighter' in Iraq and Afghanistan, China was able to stock up on new technologies that crippled the US in a theatre of operations that was arguably of greater strategic importance than Iraq and Afghanistan. Secretary Hagel thus argued in February 2014 that the new strategic reality was 'the development and proliferation of more advanced military technologies by other nations that means that we are entering an era where American dominance on the seas, in the skies, and in space can no longer be taken for granted'.[27] Hagel's own *Quadrennial Defense Review* (published in March 2014) was thus focused on 'adapt, reshape and rebalance our military' to that operational environment which in fact made the AirSea Battle analysis the basis for defence reform.[28] If one buys the convergence thesis, then building future capabilities may actually be more important than financing current operations. However, that does not make it more politically viable to deliberately underfund the wars one is actually fighting with reference to any possible future wars. The believers in the convergence thesis are thus caught in the same budgetary dilemma as the coinistas – their recipe for transformation merely adds to the defence bill in the short run, thus ruining their chances for shaping long-term transformation.

In the short run, investments in new technologies and platforms will increase procurement costs and add to the tail. In the long run, however, increased investment in convergence technologies

[26] US Department of Defense, *Quadrennial Defense Review Report*, February 2010, www.defense.gov/qdr/images/QDR_as_of_12Feb10_1000.pdf (25 September 2012), iii.

[27] Chuck Hagel, *FY15 Budget Preview*, as Delivered by Secretary of Defense Chuck Hagel, Pentagon Press Briefing Room, February 24, 2014, www.defense.gov/Speeches/Speech.aspx?SpeechID=1831 (26 February, 2014).

[28] *Quadrennial Defence Review 2014*, i.

will increase the military's teeth and lower procurement costs. The crux is the number of people employed by the armed forces – in the teeth as well as the tail. The fuel costs of the US armed forces illustrate what is at stake. Deloitte has calculated that the fuel consumed per US soldier increased by 175 per cent from the Vietnam War to the wars in Iraq and Afghanistan. In Iraq and Afghanistan, 22 gallons of fuel were consumed per deployed soldier.[29] This makes the Department of Defense the single largest institutional oil consumer in the United States. Since an F-16 burns 2,000 gallons of fuel per flight, the Air Force stands for 52 per cent of the entire US government fuel consumption, whereas the Navy stands for 33 per cent and the Army 7 per cent.[30] This demonstrates the massive costs associated with being platform-centric. The further the Department of Defense pursues convergences technologies, the more it will increase the fuel bill in absolute terms. And even if the convergence technologies can help lower fuel consumption, the need for fuel will grow in relative terms as the armed forces invest more in effects delivered by platforms rather than soldiers. Yet even if it is the F-16s that burn most of the fuel, it is the human element that adds to the costs in terms of blood as well as treasure. Deloitte demonstrates a link between the number of casualties by improvised explosive devices (IEDs) in Iraq and Afghanistan and the need to transport fuel by convoy. Furthermore, Deloitte concludes that transporting fuel and protecting fuel convoys increases the fuel costs fifteen times.[31] This is an argument for more fuel-efficient platforms and for consuming less fuel on bases, but the most effective way to reduce costs and risk is ultimately to reduce the human presence on the front line. Twenty-two gallons of fuel are saved for each soldier withdrawn from the front line in Afghanistan, Deloitte estimates. The

---

[29] Deloitte, *Energy Security: America's Best Defense. A Study of Increasing Dependence on Fossil Fuels in Wartime and its Contribution to Ever Higher Casualty Rates*, 2009, www.deloitte.com/assets/Dcom-UnitedStates/Local%20 Assets/Documents/AD/us_ad_EnergySecurity052010.pdf, 1.
[30] *Ibid.*, 10–11.   [31] *Ibid.*, 19.

cumulative effect is probably larger if one is able to reduce the need to lift and transport soldiers in the theatre.

What is true of fuel costs is true of a lot of the other central parameters in the defence budget.[32] This is one reason why the centre of attention in developing convergence technologies for the military is what the National Science Foundation terms 'uninhabited combat vehicles' and warfighter systems; laymen might call them robots. In 2003, the Defense Advanced Research Projects Agency (DARPA) commissioned a survey on the feasibility of developing and deploying robots. The survey deals specifically with bipedal, human-looking robots,[33] but the researchers' main message to DARPA was that the possibilities for autonomous systems in general were so great that the humanoid model was a limiting way of thinking. 'Robots that do not resemble Earth-type organisms may have significant military worth', the report notes, 'for example, a Centaur-type robot, with a four legged lower body and a human-type upper torso, with arms and end effectors and a head, might be very useful for military purposes'.[34] Making robots in the image of humankind is a limited way of approaching the seemingly endless possibilities for creating intelligent machines able to function more efficiently than human-based models. Still, achieving human capabilities is the first development goal. The report concludes that 'rapidly advancing technology will make militarily valuable humanoid robots feasible on the battlefield by 2015 (and as capable, efficient, and reliable as human soldiers by 2035)'.[35] Given the survey's minimum definition of a robot as something very similar to a Predator drone or iRobot's mine-clearing robot, the 2015 date holds up quite well after ten years. With regard to the

[32] Roco and Bainbridge, *Converging Technologies for Improving Human Performance*, 291.

[33] Robert Finkelstein and James Albus, *Technology Assessment of Autonomous Intelligent Bipedal and Other Legged Robots*, Final Report submitted to Dr Alan Rudolph, Defence Sciences Office, Defence Advanced Research Projects Agency, November 2004, www.robotictechnologyinc.com/images/upload/file/Humanoid%20Robot%20Final%20Report%20Rev%2013%20May%2003%20&%2021%20Nov%2004.pdf (28 August 2012), 14–16.

[34] *Ibid.*, 16.   [35] *Ibid.*, 2.

question of future development, the report identifies the greatest obstacle to the development of humanoid robots as the hardware and software for the robots to operate, and the man–machine interface to integrate them in military operations as the greatest challenge.[36] In other words, robot development depends on the realisation of convergence technologies.

Interestingly, the survey offers its own caution to its prediction when it asks military operators and robot engineers to evaluate the future application of robots. The operators are generally more sceptical and expect autonomous systems to be deployed primarily in support functions, while the robot engineers expect more widespread usage.[37] 'A typical urban combat scenario for example could have teleoperated robots up front being operated from afar and scoping out the areas', one of the robot engineers is quoted in the survey, 'backing up the robots could be foot soldiers using the information from the robots to determine where the enemy combatants are'.[38] The operators in the survey are the least concerned with the engineering issues (safety, etc.) and more concerned about the systemic effects of introducing robots. Inter-service rivalry, relations to allies and strategy doctrine are thus the highest scoring concerns among the operators.[39] Operators clearly see these innovations as being at odds with current practices and how the modern military system currently operates. The robot researchers giving statements to the survey seem to recognise this. 'I know the military would like to see one soldier operate hundreds of robots', one researcher states, 'but I think 1 soldier to 2–5 robots is more realistic since teleoperation is very important in many situations, and too difficult to automate completely'.[40] Therefore, one expert notes in the DARPA survey: 'I believe we should think in terms of the human plus a robot as a system.'[41] Such a system would be very different from the modern military system. Before exploring

---

[36] *Ibid.*, 168.   [37] *Ibid.*, 55.   [38] *Ibid.*, 182.
[39] *Ibid.*, 59.   [40] *Ibid.*, 137.   [41] *Ibid.*, 182.

the nature of this difference, however, we will focus on how the US military is investing in new technologies.

PROCURING THE FUTURE

'In the two years I was in the Pentagon', muses a former high-level Pentagon official, 'we discussed the current war for two days and future wars for the 728 remaining days'.[42] US military procurement increased on average by 8 per cent per year from 1951 to 2013. This number conceals massive differences – including a 443 per cent increase in procurement in 1951 and a 18 per cent reduction in 1994.[43] Even considering the highs and lows of the US defence budget, the fact that procurement was the largest item on the budget made the Defense Department a business focused on capital investment. As mentioned in Chapter 3, this changed in the 1990s, when operations and maintenance became the largest part of the defence budget. In one sense, this reflected a defence system more focused on the job at hand; in another sense, it reflected how the Defense Department was living on existing technologies and platforms rather than developing new ones. In many ways, the RMA debate obscured that fact in the 1990s. The RMA was about placing communication technologies from the 1990s on 1970s or 1980s platforms.

Figure 5.1 demonstrates the extent to which the US armed forces have become capital-intensive. By 1959, the Department of Defense spent more than USD 100,000 (constant) per soldier, sailor and marine as well as civilian employees. By 2002, this had doubled to USD 222,657 (constant) per person. These figures demonstrate the dramatic reduction in the number of personnel – from 4.9 million in 1952 to 2.2 million in 2002, while the defence budget has increased in real terms from USD 185,550 million in 1948 to USD 482,052 million

[42] Personal communication, August 2012. *Cf.* Aston B. Carter, 'Running the Pentagon Right: How to Get the Troops What They Need', *Foreign Affairs*, January/February 2014.

[43] *National Defense Budget Estimates for FY* 2014, Department of Defense, Office of the Under Secretary of Defense (Comptroller), May 2013 (November 2013), Table 6-8.

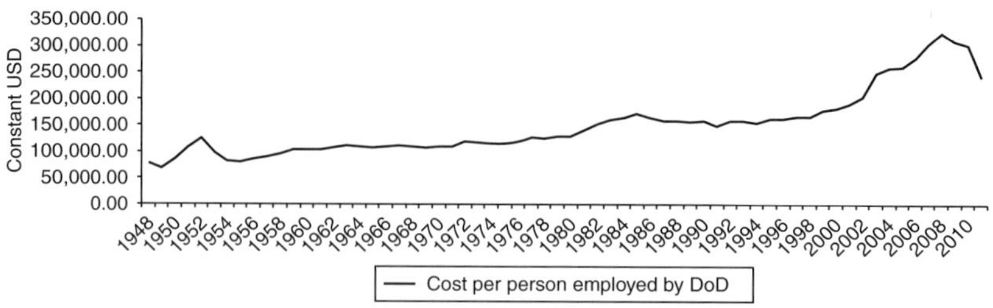

FIGURE 5.1 Cost per person employed by US Department of Defense, 1948–2011 (constant USD)

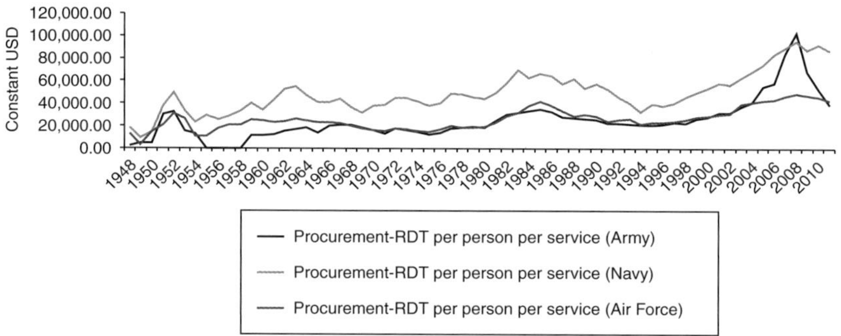

FIGURE 5.2 Procurement and R&D, training and evaluation expenditure per person per service, 1948–2011 (constant USD)

in 2002 and USD 727,966 million in 2011 (all figures constant dollars).[44] What has really happened is that the material side of the armed forces has become much more important. This becomes real if one considers the procurement and R&D part of the defence budget. Taken together, these budget titles show how much money the US has spent buying and developing materiel. Thus, it takes out pay and other issues related directly to service personnel. Figure 5.2 shows procurement and R&D in relation to the number of people – military and civilian – employed by the Department of Defense divided by service.

[44] *Ibid.*, Tables 6.3, 7.5

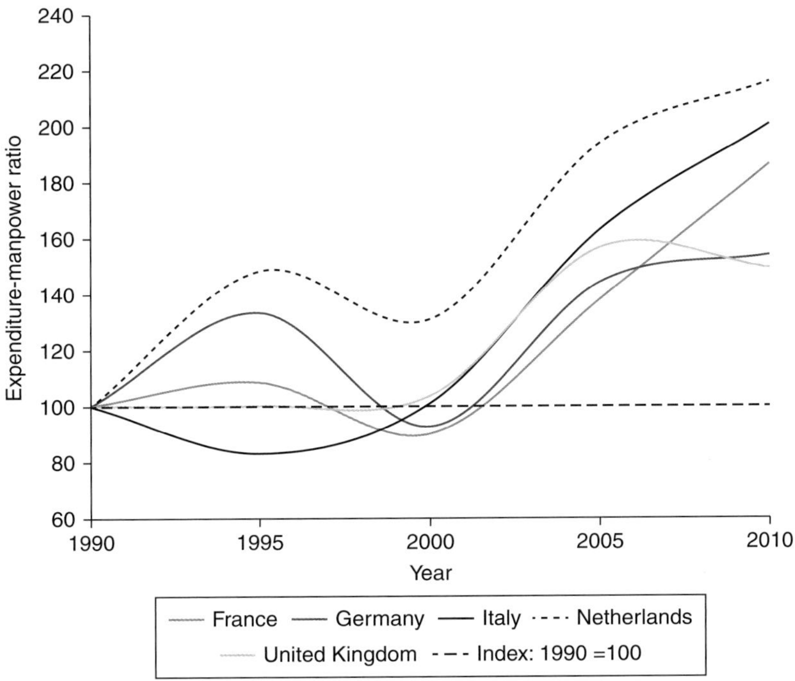

FIGURE 5.3 Ratio between total defence expenditure (constant 2010 USD) and armed forces annual strength, selected NATO countries

In constant dollars, these numbers demonstrate just how platform-focused the defence budget has become. In 1948, the Navy spent USD 18,174 (constant) per sailor and marine. In 2011 it spent USD 86,817 per employee. The Air Force spent USD 12,944 per employee in 1948 and USD 42,522 in 2011. The Army numbers are the most striking, however. In 1948, the Army spent much less per employee than the other services, just USD 1,718. In 2011, the Army spent USD 38,412 per soldier.

This is not merely a product of the US defence budget as it developed during and after the Cold War. European defence budgets show a similar tendency. Figure 5.3 shows the expenditure per soldier in larger NATO countries from 1990 to 2010. In those twenty years, the armed forces of the Netherlands more than doubled the amount spent per soldier. After a period of disinvestment immediately after the end of the Cold War, the expenditure–soldier ratio increased significantly, as operations after 9/11 and investment in new technologies drove the budgets up. Significantly, only British and German spending slowed the trend after the onset of the financial crisis. Still, this left both countries with armed forces that were 50 per cent more dependent on platforms than on people in 2010 than in 1990.

If one focuses on long-term trends, this tendency to more capital-intensive armed forces becomes even more pronounced. A comparison between the British and the US defence budget after 1945 is instructive in this respect because it shows the lack of connection between geopolitical demands on the budget and the matter of investment. While the US expanded its global commitments during the twentieth century, Britain reduced its commitments. If the defence budget is determined by power-political factors, then one would expect the British defence budget to display trends opposed to the American factors. If developments are inherent to the modern military system, however, the budget trends should be similar to those of the United States. And this is exactly what is found when studying British defence budgets over the last hundred years. In 1948, Britain spent £24,482 per person employed by the armed forces (constant pounds). In 2011, the British government paid £219,528 per person employed by the armed forces. Where the cost per person has increased by 203 per cent in the United States, the cost per person in Britain has increased by 796 per cent. The UK has maintained substantial international deployments, a nuclear deterrent and a large Navy while at the same time reducing its personnel from 830,000 in 1948 to 186,400 in 2011. These numbers are put in perspective when looking further back in the past. In 1902, Britain employed 539,000 troops to service imperial commitments and the Second South African War (1899–1902) at the cost of £20,582 per person. In 1918, Britain spent £23,328 per soldier, meaning that the UK government had to spend 47 per cent of the GDP on its armed forces, because it employed 4.5 million in the armed forces. The price per soldier thus remained relatively constant at the time when the modern military system came into being at around £20,000–25,000 per person.[45] The almost-tenfold increase in the cost per serviceperson since 1948 thus seems unrelated to the operational commitments and must primarily be determined by the increase in platform costs concurrent with the reduction in the numbers of employees in the British armed forces. One can probably also add the lack of economies of scale as the British armed forces got smaller, but in fact that only emphasises the point. Technological developments make it possible to employ fewer people, but this in turn makes armed forces more reliant on materiel.

The Western armed forces are investing heavily in materiel, and the convergence thesis makes them focus on solving the problems for the business case of the modern military system by investing in transformative technologies. In order for investment in these technologies to have a truly innovative effect, however, the convergence technologies must transform the modern military system. The next section explores some of the elements of this future business model.

THE SKINNER BOX

In the 1930s, B. F. Skinner invented what became known as the 'Skinner Box', although Skinner himself resented being turned into an eponym and preferred the term 'Operant Conditioning Chamber'. Skinner invented the chamber as part of his postgraduate studies into what conditioned human behaviour. The chamber conditions a subject – a rat, for example – to operate a device in a desired manner by giving rewards. Skinner Boxes constitute a useful analogy regarding the type of warfare convergent technologies are expected to create. This type of warfare takes the systemic approach to a completely new level, as the new depth of information and breadth to which

---

[45] 'Army cuts: how have UK armed forces personnel numbers changed over time?' *Guardian.co.uk*, www.guardian.co.uk/news/datablog/2011/sep/01/military-service-personnel-total#data (25 September 2012).

unmanned platforms can spread in the system make it possible to condition behaviour. The behaviourist approach to human action advocated by Skinner therefore also informs these new technologies of warfare. If the potential of converging technologies is realised, then the capabilities the armed forces of the future will deploy can be compared to Skinner Boxes. Regarded in this manner, it is possible to understand why so many of the misgivings about future military technologies which the West invests in seem curiously out of place in the narrative of the future as told by DARPA and others.

The 2003 National Science Foundation report was entitled *Converging Technologies for Improving Human Performance*, and improving human performance was at the heart of Skinner's project. He believed it possible to develop 'a technology of human behavior'.[46] The Skinner Box is a crude example of such a technology, which Skinner believed could be deployed in order to condition entire societies. 'As a science of behaviour adopts the strategy of physics and biology', Skinner argued, 'the autonomous agent to which behaviour has traditionally been attributed is replaced by the environment.'[47] To Skinner, it was a fact that the social environment dictated people's actions,[48] and he found it bordering on negligence that governments did not consider more carefully how they could condition individual behaviour for the common good. In so arguing, he deliberately departs from traditional notions of freedom and dignity. If you have no choice but to act well, being virtuous is a minor achievement, which was why T. S. Eliot had only contempt for 'systems so perfect that no one will need to be good'. Skinner admits that striving to be good and succeeding against various obstacles might be more impressive and commendable in moral terms, but he fails to understand why society punishes those who trespass against the boundaries of moral society while not giving direction to all people on how to behave well.[49] Skinner's concern is not individual virtue

[46] B. F. Skinner, *Beyond Freedom and Dignity* (London: Penguin, 1971), 10.
[47] *Ibid.*, 180.   [48] *Ibid.*, 178.   [49] *Ibid.*, 69.

or dignity, but the performance of the society in which the individual lives. Thus, he recognises Eliot's concerns, but merely in passing on his way to creating a better society for the future.

If human behaviour can be instrumentalised in the sense Skinner imagined possible, then there is effectively no difference between a technology shaping and directing human behaviour and a technology replicating human behaviour. In this perspective, human beings are robots, since robots would be subject to the same conditioning Skinner would apply to humans. Skinner simply does not allow for the sanctuary of a metaphysical soul or a set of biological imperatives that would make human intelligence different from artificial intelligence. Skinner can adopt this view because he has a systemic notion of the human condition. He is not concerned with specific individuals but rather with an aggregated notion of individual action, because for Skinner's boxes to succeed, not everyone has to act as conditioned; the important thing is that the majority act in the desired manner. This ensures the desired outcome and enforces the social demand of a certain type of action. It is the functionality of the system which Skinner has in mind, and human beings are merely a statistical quantity in this perspective. It is as such that they can be instrumentalised.

If one regards the military as a Skinner Box creating soldiers that are to act in accordance with a certain discipline and a specific skill set, then one is not really concerned with whether the persons involved adopt the 'Nietzschean pose' so important in the human resource perspective. A Skinner Box does not create warriors in Nietzsche's sense of empowered individuals with freedom and dignity. Coker regards this as the distinction between the free and dignified warriors and the uniform mass of soldiers. Coker uses this distinction to explain why modern soldiers find themselves alienated from high-tech warfare, and he concludes that this process of alienation will be accelerated as autonomisation sets in.[50] From a Skinner

[50] Christopher Coker, *The Warrior Ethos: Military Culture and the War on Terror* (London: Routledge, 2007), 117–31.

perspective, one could not care less. From this perspective, the only valid point is whether the instruments of the military system perform in the desired manner. If the realities of modern operations alienate soldiers to the extent they diminish performance, then there must be biochemical ways of conditioning soldiers to do their job or ways of creating machines to do the job. It is the job that matters, not striking a pose. From a warrior's perspective, war 'should be personally redemptive'.[51] If war is a defining, albeit regrettable, aspect of human existence, then it diminishes those who let machines fight their wars for them. From Skinner's instrumental perspective, these qualms may have individual relevance, but they are hardly relevant for the operation of a military system as such. Whether based on technologies directing human performance or technologies operating on their own, it is on its actions that the system should be judged – not on who performs the actions.

If the behavioural approach to warfare is pitiless with respect to human resources, it is even more utilitarian when it comes to warfare itself; that is, the operations performed by the military system. To many, this philosophical approach is a far greater problem than the military capabilities of autonomous systems. Thus, warfare by robot is often regarded as ethically inferior to types of warfare in which human beings are put at risk. An important part of this argument is risk compensation. If fighting entails little risk to oneself, then one might be more inclined to use military force. The more inclined a government becomes to use military force, the greater the risk that military force is used disproportionally.[52] This is a key element in the critique of President Obama's drone campaign in Pakistan. Critics argue that more civilians are being killed than absolutely necessary because the UAV operators in Nevada cannot possibly have a clear vision of what is taking place on the ground.

---

[51] Christopher Coker, *Barbarous Philosophers: Reflections on the Nature of War from Heraclitus to Heisenberg* (London: Hurst, 2010), 236.

[52] Mikkel Vedby Rasmussen, *The Risk Society at War* (Cambridge University Press, 2006), 74–9.

'US drone strike policies cause considerable and under-accounted-for harm to the daily lives of ordinary civilians, beyond death and physical injury', observed a report by human rights lawyers at Stanford and New York University, which concluded that 'current US targeted killings and drone strike practices undermine respect for the rule of law and international legal protections and may set dangerous precedents'.[53] The report is based on concerns for 'freedom and dignity' amongst the Pakistani population 'living under drones' in particular and concerns for human rights and international law in general. The conditioning possibly does not work, but the concept of the drone campaign in Pakistan is to apply the concept of a Skinner Box. While the US government rewards the Pakistani government with development aid and support to the Pakistani military towards stabilising their society, drones and special forces are used to punish the elements in Pakistan which the US government deems irredeemable. This introduces the notion of punishment into the behavioural approach.

Skinner notes that 'government is often defined in terms of the power to punish'.[54] From Skinner's perspective, punishment is a learning experience which functions by means of 'aversive conditioning' that makes people act in ways as to avoid further punishment.[55] The use of armed force is often described in such terms. Thus, one might argue that the United States and its allies invaded Afghanistan to punish the Taliban for 9/11. Furthermore, one could argue that the concept of AirSea Battle is to deter China from challenging US hegemony and thus functions as aversive conditioning designed to lead China to pursue peaceful integration into the international system as a great power. While punishing unwanted behaviour is a widely used technology of conditioning, Skinner finds that it is one more example of the lack of a properly refined technology of behaviour. Such a proper

---

[53] International Human Rights and Conflict Resolution Clinic (Stanford Law School) and Global Justice Clinic (NYU School of Law), *Living under Drones: Death, Injury, and Trauma to Civilians from US Drone Practices in Pakistan* (September, 2012), http://livingunderdrones.org/download-report/ (3 October 2012), vii–viii.

[54] Skinner, *Beyond Freedom and Dignity*, 64.

[55] *Ibid.*, 64–5.

technology of behaviour would 'design a world in which behaviour likely to be punished seldom or never occurs'.[56] If the United States wants to prevent al-Qaeda from attacking it or its allies or prevent China from starting a war in the South China Sea, then war has to be conceived not in terms of punishing unwanted acts but in terms of ensuring that neither al-Qaeda nor the government in Beijing acts in unwanted ways. 'It is a serious problem that we remain almost continuously at war with other nations', Skinner notes, 'but we shall not get far by attacking "the tensions which lead to war", or by appeasing warlike sprits, or by changing the minds of men (in which, UNESCO tells us, war begin). What must be changed are the circumstances under which men and nations make war.'[57]

Thomas Schelling adopted this behavioural approach to military strategy in his widely read *Arms and Influence* from 1966. '"Victory" inadequately expresses what a nation wants from its military forces', Schelling argued.[58] Punishment would not work against enemies armed with nuclear weapons or guerrillas in the mountains. Thus, Schelling focused on how to bring about a situation in which the enemy will yield to your demands without a decisive battle. Schelling believed that 'military strategy, whether we like it or not, has become the diplomacy of violence'.[59] From this perspective, victory is about forcing a decisive bargain to take place and ensuring that your bargaining position is such that you will, more or less, achieve what you want. Vietnam became the paramount example of this type of 'coercive warfare'.[60] For the very same reason, the US defeat in Vietnam discredited not only the notion of 'coercive warfare' but also the metrics used by the Robert McNamara Pentagon to measure the effects of the 'diplomacy of violence'. However, the reintroduction of the COIN argument reintroduced Schelling's argument. In 2011, British General Andrew Mackay and Steve Tatham thus published

---

[56] *Ibid.*, 69.   [57] *Ibid.*, 154.
[58] Thomas Schelling, *Arms and Influence* (New Haven, CT: Yale University Press, 1966), 31.
[59] *Ibid.*, 34.   [60] *Ibid.*, 170–8.

Behavioural Conflict, a book focusing on conditioning behaviour primarily by various information operations.[61] Furthermore, the systemic thinking inherent in the convergence view also encourages this kind of approach. In other words, the two major trends in the strategy for the armed forces focus on systemic and behavioural approaches.

A military Skinner Box will change the circumstances under which the opponent acts, thus coercing him into acting in ways more conducive to one's own agenda. Sun Tzu's notion that 'to subdue the enemy without fighting is the acme of skill'[62] would have appealed greatly to Skinner. But as so often the case with Sun Tzu's axioms, the real question is how to translate it into an operational plan. Liddell Hart's notion of an indirect approach captures Sun Tzu's sentiment, arguing that the 'true aim is not so much to seek battle as to seek a strategic situation so advantageous that if it does not of itself produce the decision, its continuation by a battle is sure to achieve this'.[63] From this perspective, seeking battle and punishing the enemy is a costly and ineffective way of achieving victory; instead, the enemy must be convinced that they will lose if they engage in battle and, equally importantly, that losing is a bearable alternative to engaging in battle due to the limited war aims of the opponent. Liddell Hart realised that by limiting your war aims, you would be able to maintain the political nature of warfare instead of relying exclusively on the fortunes of the battlefield. From this perspective, the outcome of war depends on conditioning the enemy's behaviour. In order to do so, one must create a change in how his military system operates. Changing complex systems was the concern of Donella Meadows, who studied change in large and complex systems – first as lead writer on *Limits*

[61] Andrew Mackay and Steve Tatham, *Behavioural Conflict: Why Understanding People and Their Motives Will Prove Decisive in Future Conflict* (Saffron Walden: Military Studies Press, 2011), 102. *Cf.* Rajiv Chandrasekaran, *Little America: The War Within the War for Afghanistan* (London: Bloomsbury, 2012).

[62] Sun Tzu, *The Art of War*, translated and with an introduction by Samuel B. Griffith (Oxford University Press, 1963), III, 3.

[63] Basil H. Liddell Hart, *Strategy: The Indirect Approach* (New Delhi: Natraj, 2003), 339.

to Growth for the Rome group in 1972.[64] Her approach might help us to see where the levers in the military Skinner Box should be placed. The way to change a system, according to Meadows, is to 'look for leverage points – places in the system where a small change could lead to a large shift in behaviour'.[65] This corresponds to Skinner's way of thinking. What Skinner terms 'conditioning', Meadows refers to as 'feedback loops'. Meadows notes that 'delays in feedback loops are critical determinants of system behaviour'.[66] The more slowly information travels, the less the opponent is able to make decisions. If one is able to increase the time of one's own feedback loops and delay the opponent's feedback loop, one has achieved what the US armed forces refer to as 'decision superiority'.[67] This superiority can be used to shape perceptions by denying or subverting the opponent's information. Meadows notes that this can cause systemic collapse: 'missing information flows is one of the most common causes of system malfunction'.[68] Skinner would characterise this as punishment. While punishment may serve its purposes, Skinner would recommend the creation of incentives for the opponent to act in ways that are conducive to our interests; in other words, to create circumstances where it would serve the opponent's interest to act in our interest. In Meadows' terminology, these are reinforcing feedback loops. 'Reinforcing feedback loops are sources of growth, explosion, erosion, and collapse in systems', Meadows observes.[69]

From this perspective, the Global War on Terrorism (GWT), which the United States embarked upon after 9/11, gains meaning as a transformational campaign. The invasion of Afghanistan was the initial, and largely conventional, military response to the attacks on

[64] Donella Meadows, Dennis Meadows, Jørgen Randers and William W. Behrens III, *Limits to Growth: A Report for the Club of Rome's Project on the Predicament of Mankind* (London: Signet, 1972).

[65] Donella Meadows, *Thinking in Systems* (London: Routledge, 2009) 145.

[66] *Ibid.*, 151.

[67] US Department of Defense *Quadrennial Defense Review Report*, 30 September 2001 (Washington, DC: Department of Defense, 2001), 37.

[68] Meadows, *Thinking in Systems*, 157.

[69] *Ibid.*, 155.

New York and Washington. The subsequent invasion of Iraq took the US armed forces off in the direction of developing COIN strategies. Eventually, this would lead to the call for a COIN business model for the armed forces, as described in Chapter 4. An important part of these campaigns turned out to be the intelligence-led use of special forces to take out key insurgent leaders, bomb-makers and other resources for the insurgents. These technologies could be used outside Baghdad, however. Increasingly, drones and the occasional raid became the centrepiece of the GWT, which during Obama's presidency increasingly became dissociated from the counterinsurgency campaigns. One reason for this was that the fight against al-Qaeda and similar groups no longer depended on a large conventional military presence. Where the United States had to invade Afghanistan to punish Osama bin Laden in 2001, it merely inserted SEAL Team 6 to kill bin Laden in Pakistan in 2011. In the intervening ten years, the United States developed what Mark Bowden aptly describes as a 'target engine'.[70] This is an apt description, for the United States has utilised information technology to set up a campaign model which has created a reinforcing negative feedback loop in al-Qaeda and similar organisations, the result being that the United States has gained the upper hand in the GWT. The United States is utilising analysis of massive amounts of data to map terrorist networks and identify nodes in these networks. Since terrorism is a personal business, the nodes of a terrorist organisation are radically different from those of traditional armed forces. Where the modern system is focused on the mass of the enemy, the 'target engine' seeks to identify individuals. The aim remains fundamentally the same, however: to identify the communication that makes command and control possible. The extent to which American and allied intelligence agencies seek to map global communications to identify and track terrorists became common knowledge with the revelations made by US intelligence contractor Edward Snowden in 2013. The surveillance system

---

[70] Mark Bowden, *The Finish: The Killing of Osama Bin Laden* (New York: Grove Press, 2013).

Snowden revealed is the basis for the Skinner Box. Much of the subsequent discussion about the surveillance regime focused on how this threatened the 'freedom and dignity' of individuals around the globe. From the perspective of the military Skinner Box, the issue is beyond freedom and dignity, however. Skinner's focus was not on the individual, but on the individual as a statistical value for the development of society as such. Similarly, the widespread collection of communication data serves as the basis for a 'big data' analysis. The point about analysing big data is that the large size of the data material becomes a value in itself, because even small patterns appear when the canvas is large enough. US signals intelligence does not particularly care about individual communications, but they care about the ability to create context for the individual communications that do actually prove of value for intelligence purposes.

In 2001, the United States faced the problem that the intelligence services could not clearly identify where to find the perpetrators of the attack. As President George W. Bush asserted at the time, 'this enemy hides in shadows'.[71] The rapid increase in data as well as in data analysis capability after 9/11 enabled US and allied signals intelligence agencies to prepare a much more detailed picture of the enemy. This is the foundation for creating a Skinner Box in the first place – you cannot design conditioning if you are unable to see what the rat inside the box is doing. Having identified targets, the United States was able to set the engine running. For each raid, the operators would bring back USB sticks, hard drives, papers or anything else they could find, and this would in turn generate more intelligence that pointed to new targets which would then yield more intelligence. The fuel of the target engine was thus intelligence. For example, the intelligence that ultimately led to the operation in which Osama bin Laden was killed was acquired during a raid in Iraq. Data mining combined with the 'industrial' use of special forces and

---

[71] 'Text of Bush's Act of War Statement', 12 September 2001, *BBC News*, http://news.bbc.co.uk/2/hi/americas/1540544.stm (19 November 2013).

UCAVs thus made the United States overcome the initial obstacle of finding and engaging the enemy. This led to the question of which of these targets to engage. The Obama administration designed a 'disposition matrix' that defines the parameters for which individuals should be targeted.[72] Skinner himself could have coined the concept of the disposition matrix, as it neatly summarises the belief that strategy is about conditioning subjects to act in certain predefined ways. As such, the concept also captures how knowledge production is at the centre of the target engine.

The target engine identifies leverage points – places in the system where a small change could lead to a large shift in behaviour. The real innovation is neither the UCAVs nor the special forces, but rather the insights US signals intelligence is able to get into the enemy's system, which then enables the United States to deploy assets on 'leverage points'. In practice, this has meant a campaign of targeted killings. President Obama has been at pains to emphasise, however, that the effect is much more than that. 'Much of our best counter-terrorism cooperation results in the gathering and sharing of intelligence; the arrest and prosecution of terrorists', the President noted in a speech on the future strategy of the war on terror.[73] This echoes Skinner's call for conditioning people in ways that means they will not make war. Whether this is done by arresting people before they commit an act of terror, supporting local law enforcement or providing development aid matters little, because it serves the same strategic purpose: to condition people not to fight the United States. Only when that fails are the UCAVs or SEAL teams deployed. This way of conducting military operations may well be the template for other types of military operations that rely on information and precision. This puts a premium on the military system's ability to process information. For readers of

<hr>

[72] 'Obama's Secret Kill List: The Disposition Matrix', *The Observer*, 14 July 2013; Mark Bowden, 'The Killing Machines', *The Atlantic*, September 2013.

[73] Barack Obama, *Remarks by the President at the National Defense University*, National Defense University, Fort McNair, Washington, DC, 23 May 2013, www.whitehouse.gov/the-press-office/2013/05/23/ remarks-president-national-defense-university (19 November 2013), §21.

the National Science Foundation's NBIC report, this will seem to be a precursor of how war will be fought with converging technologies. It also demonstrates the vulnerability of Western forces in such campaigns. It follows that the greatest vulnerability to a system that relies so much on information gained by signals intelligence is information itself.

In order to defend the feedback loops in one's own system, a reading of Meadows points out the creation of buffers. Buffers are large stocks with a small flow (think of a stream running from a deep lake), which make for a low average attrition rate.[74] Unmanned platforms and small nano-swarms are ideal for creating buffers, because they can be produced – and sacrificed – in large numbers. Meadows' way of thinking is close to the notion of resilience that is becoming increasingly important in Western national security discourse.[75] This is manifest in NATO's *Deterrence and Defence Posture Review* 2012, in which the Allies note that they 'are committed to providing the resources needed to ensure that NATO's overall deterrence and defence posture remains credible, flexible, resilient, and adaptable'.[76] As Steven Metz notes, 'cyberattacks might erode the traditional advantage large and rich states hold in armed conflict'.[77] The more the West depends on convergence technologies, the truer that statement will become – especially if the West deploys autonomous platforms that are wholly dependent on certain technologies to function.[78] The DARPA survey thus notes that 'counter-jamming and EMP shielding should be a major concern, as well as having a secure data link that cannot be co-opted'.[79] Western military forces can only counter that threat by being one step

---

[74] *Ibid.*, 150.

[75] Jeremy Walker and Melinda Cooper, 'Genealogies of Resilience: From Systems Ecology to the Political Economy of Crisis Adaptation', *Security Dialogue* 42 (2011), 143–60.

[76] NATO, *Deterrence and Defence Posture Review*, 20 May 2012, www.nato.int/cps/en/natolive/official_texts_87597.htm?mode=pressrelease (6 November 2012), §33.

[77] Steven Metz, *Armed Conflict in the 21st Century: The Information Revolution and Post-Modern Warfare* (Carlisle, PA: Strategic Studies Institute, US Army War College, 2000), xviii.

[78] Watts, *The Maturing Revolution in Military Affairs*, 13.

[79] Finkelstein and Albus, *Technology Assessment*, 226.

ahead of the competition. Pre-eminence thus provides a buffer for the United States. A report from the Center for a New American Security thus concludes that 'pre-eminence makes such challenges less likely and enables the United States to respond to unanticipated threats and crises, capitalize on unforeseen opportunities and preserve access to the global commons of sea, air, space and cyberspace'.[80] The problem with a very resilient system – a system with a large buffer – is that it becomes a slow and ineffective system[81] that in turn undermines the ability to act offensively on the opponent's feedback loops.

This places a premium on analysis. In their *Behavioural Conflict*, Mackay and Tatham observe that in Iraq and Afghanistan, Western forces 'oversimplified a complex system' to single factors which they could understand, but that then denied the possibility of identifying leverage points.[82] To Mackay and Tatham, that is a technocratic problem. The staff officers in Afghanistan simply did not do their job well enough, and a better understanding of human behaviour in conflict should rectify this. This reflects Skinner's notion that providing the proper technology of human behaviour can solve problems. But what if that approach is part of the problem itself? A Skinner Box presupposes the notion that one can somehow stand outside the environment in which one wants someone – e.g., a rat or person – to behave differently. This makes sense when conducting experiments on rats, but how can the conditioner be outside the society he wants to condition? Skinner's writings include many descriptions of the technologies of human behaviour, but precious little is said about what kind of behaviour one wants to further. Skinner assumes that ends are agreed upon, but, as Isaiah Berlin wisely noted, 'where ends are agreed, the only questions left are those of means, and these are not political but technical, that is to say capable of being settled by

<hr>

[80] David Barno, Nora Bensahel, Matthew Irvine and Travis Sharp, *Sustainable Pre-eminence: Reforming the US Military at a Time of Strategic Change* (Washington, DC: Center for a New American Century, 2012), 9.

[81] Meadows, *Thinking in Systems*, 150.

[82] Mackay and Tatham, *Behavioural Conflict*, 102. *Cf.* Chandrasekaran, *Little America*.

experts or machines, like arguments between engineers or doctors'.[83] In order to create a technology of human behaviour, Skinner must dismiss political and moral issues as irrelevant and frame issues of behaviour as technical questions to be discussed by social engineers. As General Omar Bradley noted in October 1971 in a speech at the Carlisle Barracks, perhaps with reference to Schelling's behaviourism as it had played out in Vietnam:

> This is the age of the computer, and if you know how to program the machine you can get quick and accurate answers. But, how can you include leadership – and morale which is affected by leadership – into your programming? Let us never forget the great importance of this element – leadership, and while we use computers for certain answers, let us not try to fight a whole war or even a single battle without giving proper consideration to the element of leadership.[84]

General Bradley is really talking about empathy; a quality that does not compute in Skinner's macro-focus on human affairs. If war becomes a matter of feedback loops and leverage points that do not take interconnected political, economic and ethical forces into account, then it is merely about acquiring targets. 'If you divorce war from all of that, it becomes a targeting exercise', General McMaster stated to the *New York Times*.

In Afghanistan, a colonel in an American headquarters experienced how information and communication technologies distanced the leadership from the realities outside the headquarters, not dissimilar to how PowerPoints were found to misshape reality by the other colonel mentioned earlier. This colonel took part in a number of conferences on whether a person identified as a Taliban operative should be killed by special forces or dealt with in a less-than-lethal

---

[83] Isaiah Berlin, 'Two Concepts of Liberty', in *The Proper Study of Mankind: An Anthology of Essays* (London: Pimlico, 1998), 191.

[84] Omar Bradley, 'Leadership', *Parameters* (Winter 2010), www.carlisle.army.mil/usawc/Parameters/Articles/2010winter/Bradley.pdf (7 June 2012), 6.

manner. His experience was that in the instances in which these conferences were done by video link, the number of people to be killed was significantly larger than when such matters were decided at regular meetings. The colonel attributed this to the more sterile communicative environment of the videoconference. One could not sense the body language of the other participants, and doubts were therefore excluded. Significantly, media such as videoconferences reinforce the premium placed on action in military systems.[85]

Basil Liddell Hart referred to this as 'scientific ruthlessness'.[86] He used the term to describe the British bombing offensive against Germany during the Second World War. 'If civilization recovers sufficiently in the years ahead for us to recover a civilized sense', Hart wrote in 1946, 'it is likely that we shall be horrified by what we have been led to do'.[87] When British soldiers arrived to occupy Germany, they were, in Max Hasting's words, 'first awed, then increasingly dismayed' by the devastation left by Bomber Command's area bombing.[88] At that point, of course, it was far too late to do anything but to pretend it had not happened. The British government never issued a campaign medal for Bomber Command (in 2013, Bomber Command veterans were offered a clasp in lieu of a medal[89]), and Commander-in-Chief of Bomber Command Sir Arthur 'Bomber' Harris was not offered a position in the RAF after the war.[90] It was not until June 2012 that the Queen unveiled a memorial for Bomber Command in Green Park. Liddell Hart's point is that the possibilities the bombers offered were applied on the basis of what this new weapon could do rather than on a strategic or moral assessment of what Britain wanted to do. This might be an argument for war being

---

[85] Personal communications, December 2011.

[86] B. H. Liddell Hart, 'War, Limited', *Harper's Magazine* 192 (March 1946), 197.

[87] *Ibid.*, 198.

[88] Max Hastings, *Bomber Command* (London: Pan, 2010), 346.

[89] 'Bomber Command Veterans Boycotting "Insulting" Award', *The Telegraph*, 17 May 2013, www.telegraph.co.uk/history/britain-at-war/10064299/Bomber-Command-veterans-boycotting-insulting-award.html (19 November 2013).

[90] Hastings, *Bomber Command* 347, 352.

too important to be left to the air marshals; because the conduct of air operations was regarded as a technocratic matter, the campaign was largely left to the RAF. 'We had a surfeit of air staffs presided over by chiefs who were not called "the air barons" for nothing', Lord Zuckerman wrote in an assessment of the bombing campaign, and continued: 'they ruled their commands like feudal allegiances. What mattered was the ability to destroy.'[91]

This ability to destroy with convergence technologies will be far greater in relation to the resources and lives one must invest in them than the aviation technologies of the 1940s. This will only make it easier to use this new tool with 'scientific ruthlessness'. Could one argue that Bill Clinton's use of cruise missiles or Barack Obama's use of drone warfare represents an example of such 'scientific ruthlessness', where a new weapons system is being continuously deployed because it seems the easier and cheaper option? President Obama certainly authorised Operation Olympic Games, which introduced malware into the Iranian nuclear facilities, in order to avoid launching a conventional attack on the Iranian nuclear programme. The President got more than he had bargained for, however, when the virus was transferred to the Internet by an Iranian scientist who probably used a USB stick to transfer data to his personal laptop and then connected to the Internet. On the Internet, the virus mutated and became known as StruxNet. 'We discussed the irony more than once', one of the President's aides commented to David Sanger of the *New York Times*.[92]

The military Skinner Box offers an illusion of control based largely on the belief that it is possible to study human beings and influence their actions without considering politics and morality; that in itself defeats the purpose of the operation, not only in moral terms, but also because it simply does not take the complicated realities of combat into consideration. It also makes strategy dependent

[91] Quoted in *Ibid.*, 350.

[92] David E. Sanger, *Confront and Conceal: Obama's Secret Wars and Surprising Use of American Power* (New York: Random House, 2012), ix–xiv.

on information systems that might not always be up to what the systems engineers promised. The New York trading firm Knight Capital experienced this first-hand. In September 2012, Knight Capital lost USD 400 million – coincidently, the price of an F-22 – in 30 minutes when its trading program malfunctioned after a software update.[93] If one imagines converging technologies malfunctioning similarly, then there is plenty to fear from the unintended consequences of NBIC weaponry; and similarly, great opportunities for those wanting to create negative feedback loops in the enemy's civilian and military systems. Bill Joy of Sun Microsystems has thus argued that NBIC may create new classes of accidents and abuses. 'We have the possibility of not just weapons of mass destruction', Joy writes, 'but of knowledge-enabled mass destruction (KMD), this destructiveness hugely amplified by the power of self-replication'.[94] I began this chapter with a dramatised description of a battle at Esfahan in Iran. What if the NBIC bombs used by the Iranians simply go out of control like Operation Olympic Games went out of control for the Americans? What if they did not stop after having subverted the American military net? One should note that information and communication technologies will probably become wearable before 2040.[95] With a brain–machine interface, the distinction between a computer and human virus will – in terms of effects, at least – be difficult to ascertain. If an NBIC bomb went viral, it would attack thousands of people and damage the worldwide communication infrastructure. Which would be more than ironic.

In sum, the convergent technologies point to a military business model that makes information much more central than in the modern military system. This greatly reduces the mass needed for combined arms operations, thus reducing the size of the military tail.

[93] 'Trading Program Ran Amok, With No "Off" Switch', New York Times, 3 August 2012, http://dealbook.nytimes.com/2012/08/03/trading-program-ran-amok-with-no-off-switch/ (13 September 2012).

[94] Bill Joy, 'Why the Future Doesn't Need Us', Wired 8(04) (April 2000), www.wired.com/wired/archive/8.04/joy.html?pg=1&topic=&topic_set= (13 September 2012).

[95] Ministry of Defence, Global Strategic Trends, 137.

The teeth of the operations will focus on the opponents' information and communication systems. Furthermore, the tip of the spear will mostly be autonomous platforms – either operating in teams with human operatives or independently. Independently operating systems might be very small indeed and use swarming techniques as nanotechnologies are fully applied in the military field. The availability of such forces will create opportunities for more holistic strategies focusing on shaping the battle space rather than seeking decisive action by massed combined arms. 'Instead of relying on massed forces and sequential operations', the Joint Chiefs of Staff's *Joint Vision 2010* states, 'we will achieve massed effects in other ways'.[96] In this section, the notion of a military Skinner Box has been used to describe how such effects can be achieved. The Skinner Box reflects the functional approach of those trying to adapt NBIC technologies to armed forces, which ought to be a cause for concern. Military aviation presents a case for how new technologies that are applied by technocratic reasoning alone can be used in ways that might realise the potential of a technology but do so in ways that neither further the strategic ends nor seem morally acceptable to the public when it realises what has actually been done. In the case of convergent technologies, the unintended consequences of use might be very profound. This will probably give rise to environmental concerns if and when convergent technologies are made operational. That time has yet to arrive, however. The concluding section deals with the question of how the armed forces are to manage the transformation from present to future capabilities.

RAPTOR SALAD

The Luftwaffe Eurofighter Typhoons had F-22 Raptor kill markings on their fuselages, and the German pilots showed them off with glee at the Farnborough Airshow, where the global aviation industry gathered

---

[96] US Joint Chiefs of Staff, *Joint Vision 2010* (Washington, DC: Chairman of the Joint Chiefs of Staff, 1997), www.dtic.mil/jv2010/jv2010.pdf, 17.

for an annual tradeshow in the summer of 2012.[97] The German pilots had not actually shot down the Raptors, of course, but the Typhoons had participated in the Alaska Red Flag exercise on Eielson Air Force Base in Alaska. The point of the Red Flag exercises is to train air combat under the most realistic conditions possible. The F-22 is the most advanced aircraft in the US inventory; so advanced, in fact, that the Americans regarded it a fifth-generation aircraft, whereas the Typhoon was deemed merely fourth generation.[98] No wonder the Germans felt that they had something to prove, and they did so the old-fashioned way by outmanoeuvring the F-22. The German pilots sped close to the F-22, thereby neutralising the Raptor's advantage in stand-off weapons, and the Germans were able to use tight manoeuvring to shoot down the Raptors, which cost around USD 400 million each. 'Yesterday, we had a Raptor salad for lunch', concluded one of the German pilots. The US Air Force was not amused and did a lot of explaining, referring to defective oxygen systems in the Raptor.[99] The problem remained that the US Air Force had not bought a technological advantage large enough to defeat a much cheaper plane with a talented pilot. On the one hand, the 'Raptor salad' story illustrates the US commitment to investing in military hardware in order be the first to use the advantages of future technologies; on the other hand, it also demonstrates the risks of that strategy.

Markides notes that top-ranked firms in a particular industry have a 91 per cent probability of survival.[100] 'Without the benefit of a technological innovation', he notes, 'it is extremely difficult for any firm to successfully attack bigger competitors or to successfully enter

[97] 'F-22 Raptor Kill Markings Shown Off by German Eurofighter Typhoons'. *The Aviationist*, 2012, http://theaviationist.com/2012/07/23/f-22-raptor-kill-markings/ (13 September 2012).

[98] Mikkel Vedby Rasmussen and Henrik Ø. Breitenbauch, *Denmark's Need for Fighter Aircraft* (Copenhagen: Danish Institute for Military Studies, 2007), 18.

[99] 'How to Defeat the Air Force's Powerful Stealth Fighter', *Wired.com*, www.wired.com/dangerroom/2012/07/f-22-germans/ (13 September 2012).

[100] Constantinos C. Markides, *Game-Changing Strategies: How to Create New Market Space in Established Industries by Breaking the Rules* (San Francisco, CA: Jossey-Bass, 2008), ix.

new markets where big established players rule'.[101] Since Marshall launched his strategy for outspending the Soviet Union in the 1970s, the United States has successfully based its security on that 91 per cent probability. This 91 per cent doctrine has depended on the US willingness to be the leader of military innovation to ensure that that last 9 per cent would not come into play due to some competitor being able to revolutionise warfare by introducing new technologies. When the United States invests heavily in NBIC technologies, it is exactly with this risk in mind. The problem for the US armed forces is that NBIC technologies are still a generation away. The DARPA survey on robotics is not exactly without enthusiasm for the subject, but even these industry experts and military robot operators do not expect fully autonomous platforms before 2035. In the meantime, the United States has an entire generation of aircrafts and other platforms procured according to Marshall's strategy in the 1970s and 1980s which are up for replacement. The F-22 and F-35 are the replacement fighters, albeit very expensive replacements. So expensive, in fact, that Secretary Gates cancelled the F-22 programme because he saw the planes as too costly. Instead, Gates wanted to focus on the multi-role F-35, which is to replace the F-16, F-18 and Harriers in the US inventory. However, the F-35 programme has had a 32.8 per cent increase in the price per unit since the project began.[102] The United States is thus investing heavily in capabilities that the armed forces themselves expect to be obsolete once the convergence technologies come online. And the United States makes this investment at such a high price that it is bound to be crowding out research and development within NBIC technologies.

The technological approach to reforming the modern military system is thus caught in the same net as the human resource (COIN) approach. The strategy for innovating the military business model

---

[101] *Ibid.*, xii.

[102] Young Hoon Kwak and Brian M. Smith, 'Managing Risks in Mega Defense Acquisition Projects: Performance, Policy, and Opportunities', *International Journal of Project Management* 27 (2009), 816.

it describes only saves money in the long run; in the short term, convergence technologies and their organisational implementation simply add to the budget. The focus on procurement is thus driving up defence budgets, not only in a superpower like the United States, but also in a mid-sized power like Britain. The fact that the UK defence budget per person has risen from £24,482 per person in 1948 to £219,528 per person in 2011 testifies to the enormous expense of wanting to maintain first-class military forces. As Augustine's law demonstrates, these costs are unsustainable in the long run. However, convergence technologies promise that these costs need not be sustained in the long run. One strategy for the Western armed forces might be to in effect skip a generation by investing heavily in the research and development of convergence technologies in the present while not developing more traditional platforms like fighter aircraft.[103] Why pay full price for the last generation of manned fighter aircraft when there are plenty of planes from the previous generation to procure? The fact that the Typhoon could outman-oeuvre the Raptor suggests that the United States could do with buy-ing F-16s and F-18s instead of replacing the entire fleet with F-22s and F-35s.[104]

By investing so heavily in last-generation platforms, the US armed forces keep the existing organisational structures in place. In many ways, AirSea Battle represents a strategy for preventing the reorganisation and refocusing of the military system advocated by the coinistas. In the short term, the focus on technology thus blocks the reorganisation of the military system which the implementa-tion of NBIC technologies would necessitate in the long run. The result of this lack of forward thinking in terms of putting together an organisation ready for the changes meriting the organisation in the first place is that NBIC issues are dealt with in a compartmentalised

---

[103] Andrew F. Krepinevich, *Defense Investment Strategies in an Uncertain World* (Washington, DC: Center for Strategic and Budgetary Assessment, 2008), 49–51.

[104] I am indebted to Henrik Ø. Breitenbauch for discussion on these issues; see also Rasmussen and Breitenbauch, *Denmark's Need for Fighter Aircraft*, 30–44.

fashion. This is demonstrated by the fact that AirSea Battle is more a shopping list than a strategy.

These mistakes do not only have consequences for the US military; every ill-conceived US procurement decision has a multiplier effect among the US allies. This is true when the United States is actively engaged in mentoring a foreign military in terms of how to conduct a war and equip it with far too expensive materiel and force structure. It is also true in the relationships with the more well-off allies. For Norway or Britain, purchasing F-35s to ensure interoperability with the United States makes sense – even if they might have been better served by procuring F-16s and UCAVs. Since Norway and Britain have a much smaller defence budget than the United States, however, the price of the F-35s will have an even greater crowding out effect on their respective defence budgets. The US commitment to buying the last-generation platforms will make it even harder for its allies to invest in convergent technologies.

The United States might be able to live with the fact that it will be the only power with a buffer in its defence budget big enough to invest in last-generation as well as next-generation capabilities. This will leave the United States with something close to a monopoly in the advanced military technology market. This secures jobs and leverage in relation to allies as well as opponents. This position also holds a huge risk, however. While 91 per cent of the dominating firms indeed remain so (see Chapter 3), according to Markides, it is also true that those who attempt to take on top-ranked firms and win do so because 'they actively adapt a different strategy (or business model) and aim to compete by changing the rules of the industry'.[105] With the United States heavily committed to procuring last-generation and next-generation platforms, the Pentagon's budget is very inflexible. This is underlined by the fact that budget cuts fall primarily within research, development and procurement. The United States thus has little choice but to reduce the force structure

---

[105] Markides, *Game-Changing Strategies*, x.

if it wants to be able to invest in future technologies.[106] The Chinese or any other competitor will be able to benefit from not being a market leader. They can learn from the Americans' mistakes, whereas the United States is so far ahead and its procurement processes and defence industry so concentrated that it has no real competitors to learn from. With that imperative efficiency of the market ruled out, the United States might make very bad and very costly procurement decisions but first notice having done so far too late. Their competitors are able to avoid repeating such mistakes. At the same time, the American dominance on the market also encourages them to follow Markides' recipe and focus on the niches the Americans choose not to explore.

In sum, making technological innovation the centrepiece of the business case of the US military comes with an enormous price tag. The business model is being implemented in a manner whereby the taxpayers must pay for the last-generation and next-generation capabilities at the same time. This may create resilience in the development process, but it does so at the risk of crowding out the niche project that is actually the real key to the future of military development. Because of the American dominance on the weapons market, however, these niches are logically the most attractive to any competitor.

[106] Frank Kendall, *Testimony Before the House Armed Services Committee*, Witness Statement of Hon. Frank Kendall, Under Secretary of Defense Acquisition, Technology & Logistics, 28 January 2014, US Congress, House of Representatives, Armed Services Committee.

6   Designer wars

The modern military system is defined by four elements: (i) combined arms operations, (ii) capital-intensive forces, (iii) focus on 'the tail' and (iv) the flow of information that allows for command and control. Like any organisational principle, the modern military system was created to meet a demand. Modern military technologies created a bloody stalemate on the western front during the First World War. The modern military system organised them into a technology of deployment that was able to break the stalemate and produce decisive results. Since the Second World War, that system has become more and more sophisticated and less and less effective. 'Simplicity and flexibility, both of them vital principles of war, were lost', as military historian Martin van Creveld observes, 'consequently, though today's Abrams and Leopards and Challengers and T72s are individually much more powerful than the Shermans and Tigers and Cruisers and T34s of World War II, as instruments of war and conquest they may be less effective'.[1] When applied in all its intricacy, the modern military system provides a dazzling display of force, as was the case with the American invasion of Iraq in 2003. Yet the very perfection to which the United States military has driven the use of the modern system demonstrates the paradox of military power. The ability of armed forces to force a decisive result in a dispute does not necessarily correspond to an ability to bring about the desired outcome. The subsequent insurgency in Iraq after the 2003 invasion testifies as much. The fact is that the modern military system seems increasingly unable to deliver decisive outcomes at the

---

[1] Martin van Creveld, *Technology and War: From 2000 BC to the Present* (New York: The Free Press, 1991), 282.

same time as the costs of maintaining the forces to operate the system are increasing. It is important to note that these costs are driven by factors inherent to the system. The tail is too large, procurement too expensive and the costs of operations and maintenance too large as well.

This book has described the nature of the modern military system and how it defines the business model for Western armed forces. We have seen how the development of the armed forces has increasingly been defined by the law of diminishing returns, which Augustine described so poignantly in his prediction that in 2040 the US armed forces will have a single aircraft to share between them. Focusing on technological innovation, an important constituency within the armed forces argues that in 2040 convergence technologies will have enabled the armed forces to escape the diminishing returns of present procurement. These new technologies will enable the armed forces to reduce the size of its tail in favour of investing in its teeth. In so doing, the modern military system will be transformed by increasing the information input into the system and thus enabling more efficient resource consumption.

The way technological possibilities shape expectations within the military business is demonstrated by a lawsuit brought against the British Ministry of Defence by the families of British soldiers killed or injured in Iraq over the alleged negligence of the Ministry of Defence to procure the materiel suitable for the mission. The Ministry of Defence had argued that the case should be dismissed on the grounds that neither the European Convention of Human Rights, which obliged the MoD not to put its employees at needless risk, nor the normal rules for duty of care applied to the battlefield, but the UK Supreme Court dismissed this argument in June 2013, clearing the way for negligence claims against the Ministry of Defence. The Court's argument demonstrates that the military is increasingly regarded on equal footing with any other public or private business. The Court recognised the risk that its ruling would contribute to 'judicialising warfare', but the judges pointed out that the risk profiles on the armed forces operations were so varied, and so many decisions on equipment etc. were taken back home in Britain in circumstances comparable with the decisions made by civilian firms or public organisations, that the circumstances for the military business were often not that unique; and, thus, the case could proceed.[2] Regardless of whether the Ministry of Defence loses the case, the fact that the government is legally responsible for providing the right kind of equipment makes it obligated to procure the most advanced type of materiel and seek the highest possible safety standards if it wants to be immune to subsequent legal challenges. From this perspective, keeping up with technological innovation is not only a matter of strategic priorities but actually a legal obligation. In that sense, the courts seem to have grasped the logic of the modern military system that is defined in many ways by the commitment to continuous technological innovation and the notion that technology, rather than strategy or tactics, is the key to military success and safe conduct on the battlefield.

The coinistas argue that this focus on technology stifles the military organisation. They want to avoid soldiers being operators that rely on the mechanics of machines and the mechanics of lawyers; instead, they want the warrior back and they find that counterinsurgency operations provide a reason why. They do so, recognising the political nature of conflict in a COIN setting, but defining the setting correctly gives them the opportunity to find a role for the warrior. Focusing on demographic trends and globalisation, the coinistas argue that investing in counterinsurgency capabilities can reverse the diminishing returns on the defence budget. Instead of the modern military system's fetish for technological innovation, they want to reframe the question of the future development of the armed forces

---

[2] *Smith and others (Appellants)* v. *The Ministry of Defence (Respondent)*; *Ellis (Respondent)* v. *The Ministry of Defence (Appellant)*; *Allbutt and others (Respondents)* v. *The Ministry of Defence (Appellant) [2013] UKSC 41*, The Supreme Court, 19 June 2013, www.supremecourt.gov.uk/decided-cases/docs/UKSC_2012_0249_PressSummary.pdf (27 November 2013).

in human resource terms. The future of the armed forces is not better platforms, but better soldiers and marines who are able to operate together with other government agencies in the highly volatile environment of the future global megacities.

In July 2013, US Defense Secretary Chuck Hagel presented the results of the *Strategic Choices and Management Review*, which were to define the Defense Department's priorities when implementing the savings forced upon the Department by US fiscal austerity measures. Secretary Hagel's statement demonstrates how the choice between a COIN model and technology-based model is widely recognised as the fundamental strategic choice for the US armed forces and, by implication, the US allies. Hagel pointed out that a reduction in Army strength and cutting Air Force platforms would enable the Department of Defense (DoD) to save USD 150 billion if supplemented by various management reforms, including downsizing headquarters. The US fiscal situation did not allow the Secretary to stop there, however. Sequestration could cut USD 500 billion from the DoD budget over a ten-year period, and, even if Congress and the President decided otherwise, this demonstrated that the Pentagon could not expect to uphold current levels of funding. Facing the budgetary facts, Secretary Hagel argued that the US armed forces had to choose between 'trading away size for high-end capability' or 'trading away high-end capability for size'. In other words, the Pentagon's review formalised the choice between what had hitherto been paradigms discussed in military circles and which this book has attempted to formulate as coherent business models. Now it was official – either the US had to focus on a business model based on technology and platforms or a business model based on deploying soldiers for COIN-style operations. 'These two approaches illustrate the difficult trade-offs and strategic choices that would face the Department in a scenario where sequester-level cuts continue', Secretary Hagel noted, clearly hoping that this stark choice would deter Congress from cutting USD 500 billion from the defence budget.

But what is interesting about the Secretary's statement is that Hagel neither rejects the existence of a fundamental choice between the two business models nor seems able to present an alternative business model that would somehow negate the dichotomy.[3] In other words, the Secretary of Defense seems unable or unwilling to present an alternative model of innovation. There are good reasons for this, since the technological and human resource approaches to innovation within the military system capture important problems and identify potential avenues of change. Neither approach has been able to deliver on the promise of innovation, however. The technological approach is perhaps best represented in the AirSea Battle doctrine. This doctrine commits the United States – and, by implication, its allies – to invest in the most recent generation technologies, such as the F-22, at a time when Western defence budgets seem unable to pay for these investments. The result is the procurement of platforms in insufficient numbers, as predicted by Augustine, and the crowding out of investments in future technologies. This is not a viable model for innovating the modern military system; nor is the human resources approach to innovation viable. The coinistas clearly identify an important driver of conflict and turn it into a driver for change within the armed forces. This is a coherent narrative that became very influential for a time, but in the final analysis, it becomes an argument for Army reform rather than defence reform. Secretary Gates acknowledged as much when he told the West Point Class of 2011 that the Army they were to serve in was not to carry out the high-end missions. The Air Force and Navy had secured these missions with the ideas that became AirSea Battle while the Army and Marine Corps were fighting in Iraq and Afghanistan. In budgetary terms, the result of the focus on COIN was to increase Army procurement for counterinsurgencies on top of the existing budget for

<hr>

[3] Hagel, Chuck, *Statement on Strategic Choices and Management Review*, as delivered by Secretary of Defense Chuck Hagel, Pentagon Press Briefing Room, 31 July 2013, www.defense.gov/speeches/speech.aspx?speechid=1798 (26 November 2013).

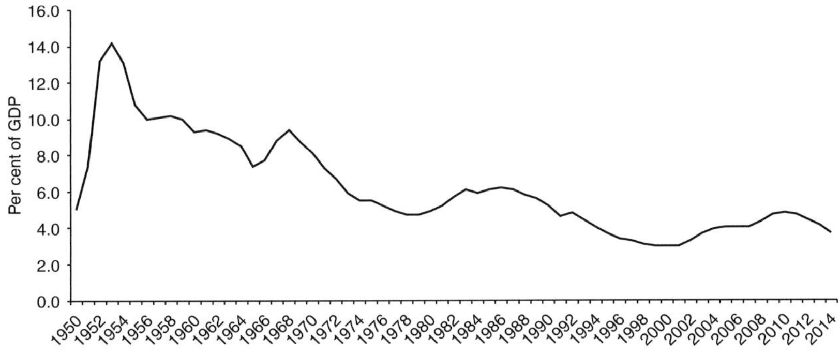

FIGURE 6.1 US national defence in percentage of GDP

high-end capabilities such as heavy armour and fighter jets. As in the case of the technological approach, the COIN approach merely adds to an already excessive bill.

For various reasons, the United States and even some of its allies might keep paying that bill. In this respect, it is telling how Britain has accepted paying almost eight times more per armed forces employee today as compared with 1948. In absolute terms, this is an excessive capital investment; in relative terms, however, it may well be regarded as a reasonable price to pay for keeping up with techno-logical development. In 1948, Britain spent 6.8 per cent of its GDP on defence; in 2011, this figure was 3 per cent. In 2011, the United States spent 4.7 per cent of its GDP on defence; in 2014, the figure is 3.7 (see Figure 6.1). The European allies are mostly paying between 1 and 2 per cent of their GDP on defence, largely ignoring NATO's 2 per cent benchmark. Even 2 per cent of GDP is a lot of money, especially in a time of economic crisis, but it is an amount a government can comfortably spend if the politicians and electorate so desire. So perhaps we should not be as interested in the willingness to pay the bill as in the point of paying it. This is why I have chosen to focus on the business case for the armed forces. If one cuts away the national, economic, social and political reasons for supporting the men and women in uniform and funding a massive defence establishment, including a large defence industry, the question remains that every high-street shop owner and every executive of a large corporation asks themselves at some point: what is our product and how do we sell it? In other words, what is our business all about?

STRATEGY MAKING

The lucidity of a strategy seems to be in reverse proportion to the number of pages spent explaining it. At least that is the impression one is left with at a time when Western governments seem none the wiser concerning the utility of their armed forces, even as they produce security strategies wholesale. In Western Europe, national security strategies are proliferating and Britain has even established a national security council to advise the Prime Minister.[4] This is yet another example of how the organisation of the modern military system in individual countries is largely shaped by American ideas about how to run the armed forces and the national security bureaucracy accompanying them. The European documents suffer from the same ailment as the American templates, however: they are strong on the description of a changing security environment and proud to list national capabilities, but weak on how to utilise national capabilities in said changing security environment. This has not gone unnoticed in policy circles on both sides of the Atlantic, and this realisation that something is amiss is the starting point for real strategy making.

The armed forces are obviously aware that the demand for their particular services is defined by global developments, and considerable effort is put into analysing these trends to provide a guide for force planning. The fact that the nature of the military system itself is taken for granted means that military scenarios tend to have an outsider's perspective on social development. As impressed as one can be with the quality of staff work on how to deploy conventional military forces, one can frankly be surprised by the lack of insight

---

[4] Henrik Ø. Breitenbauch, *Kompas og Kontrakt* (Copenhagen: Danish Institute for Military Studies, 2008).

displayed in statements like the following from the US Joint Forces Command in 2009: 'the future operating environment will be characterized by uncertainty, complexity, rapid change, and persistent conflict'.[5] Such a declaration does not leave the impression that the US Joint Forces Command has a particularly clear vision of the future security environment. In the document, the world of globalisation and rapid technological development is presented like a thousand images of change on fast-forward. The reason for this is not that the people writing the document do not have the intellectual capacity to analyse a world of transformation, but rather that the only conclusion needed for the staff work to be done on the basis of the capstone concept is the identification of change and insecurity. Knowing that the US military will be presented with a number of missions and threats that cannot be predicted, the response from the modern military system is to increase the capability to support different types of operations in different environments. Paradoxically, the notion of transformation thus becomes an argument for the status quo, because in the absence of a clear and present danger, the armed forces should retain the capacity to engage all types of contingencies. A Joint Staff analysis of the lessons of the wars in Iraq and Afghanistan thus has an insightful analysis of the types of engagement the US forces have engaged in and the demands this places on the US armed forces, but the authors write little about how the Department of Defense should operate to ensure that the United States prevails in these types of conflict.[6]

'Both public strategic documents from recent administrations and actual American strategic behaviour', Andrew Krepinevich observes, 'suggest that US political and military leaders have been increasingly inclined to equate strategy with listing desirable goals,

---

[5] US Joint Forces Command, *Capstone Concept for Joint Operations*, 15 January 2009, www.jfcom.mil/newslink/storyarchive/2009/CCJO_2009.pdf (8 November 2012), 2.

[6] US Joint Chiefs of Staff, *Decade of War, Volume I: Enduring Lessons from the Past Decade of Operations*, Joint and Coalition Operational Analysis, a division of the Joint Staff J7.

as opposed to figuring out how to achieve them'.[7] This analysis is echoed by the Public Administration Select Committee in the UK House of Commons, which concludes that 'government has lost the capacity to think strategically'.[8] In 2010–11, the committee tried to engage the government in a discussion of this matter but was disappointed by a response which, in the Committee's view, merely confirmed the government's inability to approach strategic questions as something encompassing a definition of interests and plans for carrying them out. The Committee is merciless in its dismissal of the government's response: 'the government's response confuses interests with tactics'.[9] Committee recommendations regarding the creation of a 'community of strategists' reflects the conclusion of the Chief of Defence Staff, Air Chief Marshal Sir Jock Stirrup, that 'we've been hunter/gathers of strategic talent, rather than nurturers and husbandmen', for which reason Britain had lost its 'institutionalised capacity for, and culture of, strategic thought'.[10] British experiences in Afghanistan had clearly contributed to this notion in Whitehall that perhaps it was time to start reflecting on how decisions were planned, made and implemented. The officers appearing before the House of Commons Committee to explain UK strategic planning about the engagement in Afghanistan were unequivocal in their conclusion that the deployment was not based on a coherent strategic process. A year into the campaign, a British officer enquired to regional headquarters in Afghanistan on strategy and was told 'there is no plan, Sir. We're just getting on with it.'[11] 'By definition', Rumlet

[7] Andrew F. Krepinevich and Barry D. Watts, *Regaining Strategic Competence* (Washington, DC: Center for Strategic and Budgetary Assessment, 2009), viii.

[8] House of Commons Public Administration Select Committee, *Who Does UK National Strategy? Further Report*, Sixth Report of Session 2010–11, Ordered by the House of Commons to be printed 25 January 2011, www.publications.parliament.uk/pa/cm201011/cmselect/cmpubadm/713/713.pdf (13 November 2012), 3.

[9] *Ibid.*, 3.

[10] Jock Stirrup, *Annual Chief of Defence Staff Lecture*, RUSI, Whitehall, London, 3 December 2009, www.rusi.org/events/past/ref:E4B184DB05C4E3/ (13 November 2012), §40.

[11] House of Commons Public Administration Select Committee, *Who Does UK National Strategy? First Report of Session 2010–11*, Ordered by the House of Commons to be printed 12 October 2010, www.publications.parliament.uk/pa/cm201011/cmselect/cmpubadm/435/435.pdf (13 November 2012), 11.

dryly notes, 'winging it is not a strategy'.[12] Nevertheless, the focus on initiative which is drilled into officers from the day they enter officer school to the day they retire means that there is less focus on how a job is done than on getting the job done. This is not to say that initiative is not a crucial part of front-line leadership, nor is it to say that the modern military is not focused on procedures and analysis. The results are open to interpretation, however, and that makes it difficult to assess the processes leading to a given result. For example, it is in the interests of the NATO armed forces engaged in Afghanistan to define the conflict there as a success, but in defining the engagement in this way there is little reason to delve into how the campaign is conducted. This way of thinking stands in the way of distinguishing between intent, process and result. The nature of conflict involves a risk of defeat, and defeat does not necessarily mean that the vanquished army did a bad job; if that were so, the *Wehrmacht* would have prevailed in the Second World War. Only by declaring the engagement in Afghanistan a failure can genuine, serious discussion about the conduct of the campaign begin.

Strategy is about intent as well as the processes establishing and implementing that intent. From this perspective, the most important recommendation from the UK Chief of Defence as well as the House of Commons report is the need to nurture a strategic community – within as well as outside of the government – that is able to sustain a discussion about strategy. In the absence of such an ongoing debate, strategic discussion has a tendency to get stuck at first principles. In such cases, the academics in the group end up spending everyone's time discussing the impact of globalisation, the changing nature of armed conflict or the decline of American power. As the civil servants and officers present soon realise, these discussions are the antithesis of strategy because they remain abstract and theoretical, whereas strategic is concrete and analytical. Air Chief Marshal Sir Jock Stirrup expressed this very sentiment when he wondered what concepts such as 'asymmetrical conflict' and 'hybrid warfare'

---

[12] Richard Rumlet, *Good Strategy/Bad Strategy: The Difference and Why it Matters* (London: Profile Books, 2011), 128.

tell him about how to go about his business. 'All too often', Stirrup concludes, 'those very models that are supposed to help us become prisons, rather than prisms'.[13] Actually, Stirrup's critique cuts two ways: the Air Marshal challenges academics to engage with the concrete problems rather than invent abstract concepts, but he also challenges his fellow officers not to take refuge in abstract concepts. Appearing before the Senate's Armed Services committee to be confirmed as the Army's new Chief of Staff, General Martin Dempsey noted that 'we generally believe that the future will be more a series of hybrid threats, where you have to be prepared to confront your adversary wherever he chooses to confront you'.[14] In this context, the notion of hybrid warfare, which on its own terms is a creative piece of scholarship, becomes almost meaningless. The modern military system is the underlying cause of this. As long as the armed forces are regarded as a closed system which must be applied to a particular problem, then concepts such as hybrid warfare just add to the target list rather than change how the armed forces are organised, which is the real implication of the concept. From this perspective, Rumlet observes that 'a great deal of strategy work is trying to figure out what is going on'.[15] AirSea Battle offers a good example of this. It is basically a list of objectives and a corresponding list of capabilities required to reach each objective. In short, it is a shopping list. 'A long list of "things to do", often mislabelled as "strategies" or "objectives" is not a strategy', Rumlet dryly notes, 'it is just a list of things to do'.[16] In times of austerity, such lists are more difficult to pass for strategies, which is perhaps a real opportunity. As Rumlet notes, 'strategy is scarcity's child'.[17] A government department does not earn its own

<hr>

[13] Stirrup, *Annual Chief of Defence Staff Lecture*, §46.

[14] US Senate Hearing to Consider the Nomination of General Martin E. Dempsey, USA, For Reappointment to the Grade of General and to Be Chief of Staff, United States Army, www.armed-services.senate.gov/Transcripts/2011/03%20 March/11-08%20-%203-3-11.pdf (15 November 2012), 20. *Cf. Quadrennial Defence Review 2014*, vii.

[15] Rumlet, *Good Strategy/Bad Strategy*, 79.

[16] *Ibid.*, 53.   [17] *Ibid.*, 62.

money and is therefore exclusively focused on spending the taxpayers' money. Since the funding that the department is allocated is a reflection of government priorities, it is perfectly legitimate to focus on how to spend the money allocated. If a department of defence is entrusted with securing the nation, then perhaps it is understandable that it will present long lists of objectives and capabilities in order to be able to fulfil that duty under any circumstance. When public funds are scarcer, however, opportunity opens up to approach strategy differently. It is no coincidence that the British and American reflections on strategy come at a time of profound economic crisis.

STRATEGY DESIGN

The pieces of the puzzle appear to be there. The armed forces are analysing the security environment, civil servants are working to restructure the budgets, and academics are producing concepts to describe the future of armed conflict. None of these pieces seem to fit together in a puzzle that defines a future for the armed forces beyond the modern military system, however. The fact that so many people work vigorously to produce strategy without comprehensively addressing core issues seems to suggest that the entire approach to strategy-making has room for improvement. If one reduces the problems of the Western armed forces in the same manner as a chef reduces sauces, then the problem boils down to how the armed forces spend money. Most accounts – official and non-official alike – of the present predicaments of the Western armed forces begin with the reduction of defence budgets. The fact that defence spending is being cut – and will probably be cut further – by many Western governments is not the problem, however. The problem is that the strategies for using the defence budget are such that the armed forces can actually only just operate, let alone innovate, by increasing budgets. The result is a shift in focus from how the system operates to how one can operate the system and tweak it to produce; hence, the focus on abstract concepts, future technologies or operational concepts. There is nothing wrong with focusing on these issues, but they are no substitute for engaging with the elephant in the room: how the modern military system is delivering less and less for more and more money.

The military's problem is that it takes its point of departure in the system it *has* rather than the system it *needs*. Brian Leavy explains how two types of processes operate within the organisation: the first is governed by a rationality of reliability, where the production of predictable outcomes is the most important thing. An automobile factory wants the quality of the cars it produces to be uniform and the military wants to be sure that staff officers produce battle plans that can be executed in keeping with how the forces have been trained. Another type of process is governed by a rationality of validity. From this perspective, the focus is on producing better cars and more imaginative battle plans. Most executives prefer a balance between reliability and value.[18] However, government is often ruled by a logic of reliability. Civil servants are to carry out the wishes of the executive or legislature, which leaves little room for validity. As Frank Ostroff points out, this logic of reliability shapes the careers of most government leaders, including military officers. They are promoted because of the system not against it. Furthermore, that system is entrenched in rules and regulations regarding procurement, personnel, budget and other crucial factors for change – all of which are subject to public debate and powerful lobbies.[19] Adding to this mix the fact that the modern military system places a premium on reliability in order for it to apply armed force to the centre of gravity, one understands better why the system is being taken for granted. When operating in terms of reliability, the only response to new challenges is to trim the system. In Europe, the result of defence cuts is therefore ever smaller versions of a modern military establishment complete with navies, air forces and armies, the size of which seem ever

---

[18] Brian Leavy, 'Design Thinking: A New Mental Model for Value Innovation', *Strategy and Leadership* 38 (2010), 9. See also, Roger Martin, *The Design of Business: Why Design Thinking is the Next Competitive Advantage* (Boston, MA: Harvard Business Press, 2009).

[19] Frank Ostroff, 'Change Management in Government', *Harvard Business Review* May (2006), 142.

more futile from an American perspective. In Christian Mölling's apt term, the European countries are in the process of creating 'bonsai armies'.[20] Augustine would recognise this process. The military system produces reliable outcomes, even if these outcomes make sense only to the system itself.

The armed forces must be guided by a process of validity focusing on how the 'military product' can be innovated to create value in the twenty-first-century security environment. Focusing on value production will also make it much easier to prioritise the resources allocated to defence. In 2010, the US Defense Business Board concluded that the 'DoD lacks detailed, consistent and accurate cost data for major elements of its enterprise, and the different categories of overhead-type activities such as base operations, headquarters and logistics'.[21] From a business perspective, this was a very sloppy way of spending resources, but for a military system focused on reliability, redundant systems were more important than optimising the costs in relation to production. There was a limit to the optimisation that could be achieved by adopting a logic of validity in the Department of Defense's work practices. If the guiding question for a reliability-based approach is 'How are we certain to deliver?', the guiding question for a validity-based approach is 'How can we provide new options?' Approaching the questions we have covered thus far in this value-frame of mind, one ends up with a set of recommendations looking something like this:

- Regard the military as part of the national security capacity.
- Use technology to make first-generation NBIC platforms rather than last-generation conventional platforms.

---

[20] Christian Mölling, *Europe Without Defence*, SWP Comments 38, November 2011, Stiftung Wissenschaft und Politik, Berlin, www.swp-berlin.org/fileadmin/contents/products/comments/2011C38_mlg_ks.pdf (13 November 2012), 3.

[21] Defense Business Board, *Managing DoD Under Sustained Topline Pressure*, FY10-2, January 2010, www.dtic.mil/cgi-bin/GetTRDoc?AD=ADA556844 (30 January 2014), 4.

- Use the transformation of the military from a service to a profession of arms as an opportunity to transcend the dichotomy between warrior and soldier.
- Utilise global networks.
- Respect the paradox of military power.

Regard the military as part of the national security capacity

Regarding the military as a part of the national security capacity is an important first step in a more value-oriented approach to armed force. From this perspective, armed forces deliver decisive outcomes in a restricted time and space in circumstances in which the use of military capabilities is the only policy option left to a government. This is a definition of a business model that addresses the armed forces' deliverables very differently from how the modern military system has done. It seeks to identify the unique capabilities that enable the military to force a solution to a situation that cannot be solved in any other way. However, it recognises the paradox of military power to the extent that the military should be focused on delivering the best possible military solutions as part of the entire government's efforts in a certain area. Moreover, this definition recognises the limited time and space within which military power is actually powerful.

The notion that the armed forces are designed to deliver decisive outcomes in a restricted time and space in circumstances in which the use of military capabilities is the only policy option left to a government thus leaves out a number of traditional considerations. It is important to note that this is not a definition of war, so it does not include the fact that warfare consists of destroying things and killing people, nor does it address the urgent moral and political questions following from that fact. National security strategies seek to address the political questions; there is a possibility that such documents can address the moral questions of warfare more coherently – including the questions not directly related to a government's liability under international law. These types of considerations are important for how you conduct your business and are increasingly

part of the strategic process of private businesses. In the final analysis, however, they are an add-on to what Rumlet describes as the
kernel of good strategy, which is a diagnosis of the circumstances for
action that can guide policy and coherent action.[22]

The coinistas have correctly called attention to the political,
economic and social context of military operations. For practical reasons, the military has often been responsible for securing the link
between civilian and military efforts. More often than not, the net
result of this has unfortunately been a militarisation of, for example,
development aid, which was actually detrimental for the objective
of the mission. In Iraq and Afghanistan, militaries have been adept
at portraying the lack of success as the result of a failure by civilian
agencies to back up the military effort. As Karl Eikenberry points
out in the case of Afghanistan, however, the US military was able
to define the entire effort on its own terms, and in many instances
took it upon itself to deliver 'civilian' efforts (including teaching
civil–military relations) in order to ensure that the job was done in
accordance with the COIN doctrine.[23] Rajiv Chandrasekaran concludes that neither the military nor the civilian agencies were able
to translate the doctrines into practice on the ground.[24] According
to *The Guardian*, a British government report concluded that the
focus on building schools, roads and other infrastructure projects in
Afghanistan's Helmand Province as a part of the COIN effort had
produced a 'mismatch between the value of the assets and the Afghan
government's ability to maintain them'.[25] This is civil-servant speak
for failure to implement the COIN manual's rhetoric on integrating
civilian and military assets. A substantial part of that failure stems

<hr>

22 Rumlet, *Good Strategy/Bad Strategy*, 7.
23 Eikenberry, 'The Limits of Counterinsurgency Doctrine in Afghanistan', Foreign
   Affairs, September/October 2013.
24 Rajiv Chandrasekaran, *Little America: The War Within the War for Afghanistan*
   (London: Bloomsbury, 2012).
25 'Afghan Schools and Clinics Built by British Military Forced to Close', *The
   Guardian*, www.guardian.co.uk/world/2012/sep/27/afghan-schools-clinics-
   built-british-close?intcmp=239 (6 November 2012).

from how the military system works independently from other government agencies.

Military capabilities have to be regarded as part of the security capability of an entire state. The security capability includes the resilience of civil society and its ability to contribute to the nation's security as well as the ability of government departments – from the intelligence services to the ministry of commerce – to work together. The military effort should been seen as an element of such an aggregated effort. In theoretical terms, most students of strategy would regard this as pretty basic stuff. Liddell Hart called it 'grand strategy' and observes that 'fighting power is but one of the instruments of grand strategy – which should take account of and apply the power of financial pressure, of diplomatic pressure, of commercial pressure, and, not least of ethical pressure, to weaken the opponent's will'.[26] Liddell Hart realised that if aggregated capabilities are not merely counted for the benefit of league tables of national power but actually put to coordinated use, then not only will a government be able to operate more effectively but the armed forces will also be able to define their mission more clearly. For our purposes, it is crucial that the armed forces have a clear strategy that translates into clear budgetary goals, but that aim can only be achieved if what the armed forces should be doing – as well as what they should *not* be doing – is well defined. Identifying what they should be doing will dramatically increase the chances of creating what Barno terms 'aggregate capability' within the armed forces themselves by creating a greater total effect by utilising the 'the synergy of all military systems working together'.[27] In sum, this approach focuses on creating synergies between the military and other parts of government as well

---

[26] Basil H. Liddell Hart, *Strategy: The Indirect Approach* (New Delhi: Natraj, 2003), 336.

[27] David Barno, Nora Bensahel, Matthew Irvine and Travis Sharp, *Sustainable Pre-eminence: Reforming the US Military at a Time of Strategic Change* (Washington, DC: Center for a New American Century, 2012), 17.

as creating synergies within the military itself. Joint Chief of Staff General Dempsey recognised as much, concluding that

> our aim should be a versatile, responsive, and decisive Joint Force that is also affordable. This Joint Force must excel at many missions while continuously adapting to changing circumstances. It means building and presenting forces that can be moulded to context – not just by adding and subtracting, but by leaders combining capabilities in innovative ways.[28]

If capabilities are to be combined in innovative ways, then the way the armed forces coordinate services and functions as well as the relationship between the armed forces and the rest of society's capabilities must be defined in terms of what they can achieve together. This has to be defined in terms of synergies. Defining the military business model in terms of synergies involves a very different approach to strategy than the modern military system's way of thinking. The modern military system identified victory in terms of creating a closed system that ensured the application of combined arms at the right place at the right time. The information and logistical technologies of the early twentieth century were such that this could only be achieved by creating huge redundancies in the system in order to facilitate command and control as well as the timely execution of operations. In other words, it created an ever larger tail. In turn, this meant that military strategy was defined in terms of when and where to deploy the military system, whereas the 'how' was largely given by the logic inherent in the system itself. This also meant that fighting a conflict was left largely to the military to avoid interfering with the complicated deployment of the military system.

The focus on synergies transcends systemic boundaries. This is reflected in a number of current writings on defence and security. As the basis of reducing the British defence budget in 2010, the

---

[28] US Joint Chiefs of Staff, *Chairman's Direction to the Joint Force*, 6 February 2012, http://usacac.army.mil/cac2/repository/020312135111_CJCS_Strategic_Direction_to_the_Joint_Force_6_Feb_2012.pdf (8 November 2012), 7.

government undertook an analysis that included the entire government security sector in order to identify synergies. *The Strategic Security and Defence Review* was thus focused on identifying feedback loops in the British system that could either reduce the effect of cuts or identify where not to cut in order to reduce British capabilities more than cutting a certain budget title would in itself suggest. The bureaucratic politics of the actual defence cuts might have worked somewhat against this holistic approach, but the point of departure was not the shopping-list approach to strategy.[29] Captain Wayne Porter of the US Navy and Colonel Mark Mykleby of the Marine Corps are advocating the same approach in the United States. 'We must recognize that security means more than defense and sustaining security requires adaptation and evolution, the leverage of converging interests and interdependencies.'[30] In order to utilise these synergies, Porter and Mykleby similarly argue that the United States needs a coherent national security narrative. The British review and American report focus on the same deep insight: because the conditions of the moment will inevitably shape new strategies for specific issues, the point is not to decide a new strategy; rather, the point is to create a coherent strategic process that enables present and future decisions to be taken within a given framework that furthers a certain set of aims. Rumlet summarises this way of thinking: 'many effective strategies are more designs than decisions'.[31]

The use of armed force must be planned and executed, not in terms of what the military can do, but in terms of what armed force can be used to achieve as a part of a national effort. Only then can the military stop taking how the modern military system defines the

---

[29] *Securing Britain in an Age of Uncertainty: The Strategic Defence and Security Review*, Presented to Parliament by the Prime Minister by Command of Her Majesty, October 2010, http://webarchive.nationalarchives.gov.uk/20121015000000/ http://www.direct.gov.uk/prod_consum_dg/groups/dg_digitalassets/@dg/@en/ documents/digitalasset/dg_191634.pdf (29 November 2014).

[30] Wayne Porter and Mark Mykleby, *A National Strategic Narrative*, Woodrow Wilson Center, 2011, www.wilsoncenter.org/sites/default/files/A%20 National%20Strategic%20Narrative.pdf (30 September 2012), 5.

[31] Rumlet, *Good Strategy/Bad Strategy*, 129.

use of armed force for granted. The military is by no means alone in making the mistake of taking how you do business for granted. Large companies make the same mistake. By the 1990s, IBM had gone from being one of the world's most successful technology companies to being on the brink of bankruptcy. Lou Gerstner was appointed CEO by a board which knew this was probably the last chance for IBM. Gerstner realised that the company was more focused on what the employees wanted to produce than with what the customers wanted to buy. He began Operation Bear Hug, which involved Gerstner sending the top fifty IBM executives to visit customers in order to acquaint themselves with their needs.[32] IBM expanded on this notion of cooperating with the customers when the company acquired the PwC consulting firm in 2002. IBM was not merely in business to sell computers; the company wanted to sell their customers the future. This was the basic idea of the company's 'On Demand' strategy that was to make IBM a consultancy as well as a producer of computers. IBM invited key customers to take part in discussing future demands, and on the basis of the scenarios which they developed together, IBM was able to meet demands before the customers came asking or perhaps even realised what they needed.[33] The IBM approach to transformation thus differed radically from the one displayed by the US Joint Forces Command. To IBM, the future was not an unknown entity; instead, the future could be shaped on the basis of a vision of what one wanted to achieve. In that sense, IBM was able to make transformation a strategic asset rather than a trend that gradually crippled the company.

IBM's approach to future challenges demonstrates that technological innovation is as much about the 'performance attributes' as about the technology itself. 'Performance attributes' is Joseph Bower and Clayton Christensen's term for the new organisational

[32] Yves Doz and Mikko Kosonen, *Fast Strategy: How Strategic Agility Will Help You Stay Ahead of the Game* (London: Pearson, 2008), 37.
[33] *Ibid.*, 38.

possibilities and new activities created by technological innovation.[34] From this perspective, the most important aspect of technological innovation is how new capabilities are utilised, because technologies are often disruptive with regard to the present organisation and activities. Bower and Christensen point out that disruptive technologies often perform worse than existing technologies on one or more of the parameters currently valued by firms in a given market. They do new things, but not necessarily the old, valued stuff as well.[35] Applying new technologies effectively therefore means that one must redefine what the organisation does. IBM's brilliant solution to this was setting off to discover the new possibilities of technologies together with their customers, thus making the creation of new products an integrated part of organisational change. Because the armed forces maintain a rigid distinction between the different military functions as well as between the military and civilian systems, the disruptive nature of technological innovation is made more profound than it has to be. Directed-energy weapons might not be able to do exactly what artillery currently does, and that will create misgivings among artillery regiments, just as the use of drones challenges the future careers of Air Force pilots. Yet these new technologies open new ways of conducting security policy as well as the prospect of considerable savings once implemented.

One of the merits of the coinistas' argument is their attention to what the customer wanted. If the Bush administration really wanted to be engaged in Iraq, then General Petraeus could offer the way to do it. It is worth noting that the coinistas' recipe for how to prevail in Iraq was not initially received well by the administration, so this was not a case of 'the customer is always right', but rather a professional military approach to finding a way to deliver on a political ambition. However, the customer might not necessarily be right even if they get their way. Bower and Christensen note that big companies fail

---

[34] Joseph L. Bower and Clayton M. Christensen, 'Disruptive Technologies: Catching the Wave', *Harvard Business Review* January–February (1995), 44.

[35] *Ibid.*, 45.

to innovate because they follow the management textbook and stay close to their customers. 'IBM's large commercial, government, and industrial customers saw no immediate use for minicomputers', Bower and Christensen observe, and IBM therefore saw little commercial incentive to develop technologies that would almost put it out of business two decades later.[36] As mentioned, IBM tried to circumvent this problem by engaging in scenario development together with their customer. They provided the professional input necessary for customers to be able to think ahead and the company was therefore able to institutionalise the forging of alliances between the development-minded in IBM and their customers, thereby ensuring that neither IBM nor its customers stifled innovation. Such coalitions for change are vital if the armed forces are to be able to embrace future opportunities rather than being baffled by the range of possibilities. The task is to create the conditions for 'anticipatory governance' by integrating foresight into decision-making processes which are themselves to be better networked in a whole-of-government approach.[37] Leon Fuerth argues that this is the key to ensure 'actionability'.[38]

Yzes Doz and Mikko Kosonen argue that Lou Gerstner made IBM a 'dancing elephant'.[39] For the military to learn those steps, officers will have to embrace their customers in the way IBM's executives did. Yet the embrace of customer demand was one of the points where the COIN doctrine has been criticised the most. Many in the US military found that the COIN doctrine eschewed the force into a counterinsurgency focus that was far from relevant in other scenarios for military deployment. If one subscribes to a reliability-based focus on the ability of the armed forces to respond to every possible contingency, then this was a logical argument; and it followed that after Iraq and Afghanistan, the force should be rebalanced. This seems to confirm Samuel Huntington's

---

[36] *Ibid.*, 44.
[37] Leon S. Fuerth, *Anticipatory Governance: Practical Upgrades*, forwardengagement.org, 2011, www.forwardengagement.org, (22 November 2012), 4.
[38] *Ibid.*, 13.
[39] Doz and Kosonen, *Fast Strategy*, 36.

observation that 'war is more unsettling to military institutions than to any others'.[40] When the troops are actually engaged in a campaign, the carefully planned peacetime schedule is disrupted entirely, and delicate compromises on the budget are blown to pieces by operational necessity. This explains officers' preference for 'balance' in the force structure, but it is not an argument for balance in itself. In fact, the argument for balance is an argument based on internal necessities rather than external demands. It ignores the actual priorities of politicians and need of the warfighter in favour of abstract notions of what the armed forces ought to be able to do. At the heart of the notion of a balanced force is the expectation that real war is something akin to the world wars that gave birth to the modern military system.

This is by no means unreasonable. Any organisation would like to measure itself in relation to when it is operating at its full potential. The manager of a Starbucks wants to know whether her staff can deal with the morning influx of customers, not whether the staff can make a few lattes 'to go' in the quiet hours of the afternoon. By the same token, generals want to prepare their soldiers for large-scale combined arms operations. But whereas the Starbucks staff has to make coffee every morning seven days a week, the generation of soldiers training for combined arms manoeuvres at Fort Hood or on Salisbury Plain might never be deployed for such operations. This is a problem inherent in military training and planning, but while recognising that fact one should also recognise that the important competence to teach soldiers is the ability to adapt to changing circumstances. 'The US military has honed its warfighting skills based on a narrow set of scenarios that envisaged them either repelling or conducting large-scale land invasions', Jim Thomas from the Center for Strategic and Budgetary Assessment (CSBA) told Congress in 2011. In Thomas' view, this focus on large-scale conventional battles has led to 'over-optimising forces for repelling invasions and conducting

---

[40] Samuel Huntington, *The Soldier and the State: The Theory and Politics of Civil–Military Relations* (Cambridge, MA: Belknap Press, 1964), 69.

counter-offensive, large-scale land invasions in kind', thus leaving the US armed forces less prepared for a host of other – arguably more probable – scenarios.[41]

The US armed forces set the Western standard for the training and planning of future operations – formally in NATO and on a bilateral basis as well as informally by inspiring smaller Western armed forces in how to operate. This means that the needle on the Western military compass is pointing towards large-scale combined arms operations on the divisional scale experienced in the world wars. From time to time, current operations might change the course, as in the case of the wars in Iraq and Afghanistan, but the training regime will ensure that attention again points towards large-scale land invasions. This reduces what Keith Shimko refers to as the 'fungibility' of military forces.[42] Believing that large-scale combined arms operations are what the military business is fundamentally about, the armed forces have regarded any other type of operation as a lesser operation that could somehow be included in the training regime that would make soldiers ready for large-scale combined arms operations. As Shimko politely notes, 'the military's experience in Iraq, however, has led many to rethink this traditional view'.[43] The heavy modern military system does not do counterinsurgency very well, nor does it do other operations which are less conventional well. This should come as no surprise. What is surprising is the insistence that training in large-scale combined arms operations will qualify for doing these other types of operations. Perhaps one reason is that if other types of operations cannot be subsumed under conventional operations and if these conventional operations have been less sought after by the military's 'customer', then the implication is a wholesale reassessment of what the military business actually entails. This would always be a

---

[41] Jim Thomas, *Department of Defense Investment in Technology and Capability to Meet Emerging Threats*, Testimony Before the US House of Representatives, (Washington, DC: Center for Strategic and Budgetary Assessment, 2011), 3.

[42] Keith L. Shimko, *The Iraq Wars and America's Military Revolution* (Cambridge University Press, 2010), 226–9.

[43] *Ibid.*, 227.

difficult question to ask in any organisation, but this question is all the more problematic in the armed forces, because the modern system has only offered one answer. Substituting one answer with several demands a new way of engaging with the system's environment. This is perhaps the fundamental reason why future developments are often presented as a blur of change in military scenarios. In the final analysis, they are distractions from real warfare.

IBM did the exact opposite by focusing on what the company needed to be able to deliver in the future. IBM was not supposed to be in 'balance' but rather it constantly repositioned itself in relation to future demands. The notion of armed force as a part of a national security capacity focuses the military on creating value for a coherent government effort rather than supplying a reliable, balanced force. An unbalanced force should constantly adapt to new challenges and technologies. In fact, this is a rather fitting description of how the US military and its Western allies have been changing constantly since the end of the Cold War. Military circles focused on reliability have seen this as a problem, but the real trick might be accepting this as a condition for the military business in the twenty-first century.

Use technology to make first-generation NBIC platforms rather than last-generation conventional platforms

Embracing the notion of creating value by embracing change is crucial for how Western armed forces are approaching procurement. If the military is to follow IBM's lead and become a 'dancing elephant', then it has to embrace the NBIC revolution. First of all, it has to do so because the utility of existing platforms is being challenged. In July 2011, Jim Thomas of the CSBA testified to Congress that 'American power projection in its familiar form could become prohibitively costly in the future as A2/AD battle networks proliferate.'[44] AirSea Battle is meant to address this issue, but it is more likely to merely

---

[44] Thomas, *Department of Defense Investment in Technology and Capability to Meet Emerging Threats*, 3.

intensify the arms race between offensive and defensive weaponry in the Western Pacific. The offence is unlikely to win this race, but by investing in offensive, access technologies, the United States and its allies might end up procuring last-generation platforms at a rate that makes investing in future NBIC technologies prohibitively expensive. AirSea Battle is a recipe for ensuring reliability at a time when the development of NBIC technologies puts a premium on creating new value. By pursuing last-generation platforms by means of AirSea Battle, the United States and its allies risk making the mistake IBM made in the 1980s by investing in the technologies that were 'right for today but wrong for tomorrow'.

Like IBM, the armed forces must think about how to utilise new technologies together with the 'customers' in order to develop products that are relevant for the future national security strategy. Doing so is difficult, because NBIC technologies fundamentally challenge how the Department of Defense and its European counterparts go about their business. Andrew Krepinevich notes that 'today's Department of Defense planners have little experience in crafting investment strategies during periods of military discontinuity'.[45] Military discontinuity is nonetheless going to be a defining feature of the years to come due to technological development and because discontinuities are going to create value for the defence departments of the future. From this perspective, the best service the civilian part of government could do the military is to be a partner in a structured dialogue on future requirements in the way IBM and its customers engaged one another in defining future business technologies. This can only be achieved by creating or re-creating a strategic community which can engage in such discussions.

At the heart of these discussions will be the extent to which NBIC technologies are able to provide the buffers that have previously been supplied by conventional forces. The notion that, in spite

---

[45] Andrew F. Krepinevich, *Defense Investment Strategies in an Uncertain World* (Washington, DC: Center for Strategic and Budgetary Assessment, 2008), 7.

of being of little immediate relevance, it was nevertheless most prudent to prepare for large-scale combined arms operations is a sound enough premise; that if it came down to a large-scale conventional confrontation, it would be impossible to supply an adequate force on the spur of the moment. Awareness of this inability would in turn cripple national policy and encourage the nation's enemy. From this perspective, national defence is an insurance policy and, as in the case of fire insurance, one should count oneself lucky if one never has the occasion to make a claim. In that sense, defence is what Meadows would describe as a 'buffer', of which technology is a very important part. 'In modern war', Adam Smith argued in the *Wealth of Nations*, 'the great expence [sic] of fire-arms gives an evident advantage to the nation which can best afford that expence; and consequently, to an opulent and civilized, over a poor and barbarous nation'.[46] Smith was concerned with the permanence of modern society – would modern British society collapse in the same way Rome had collapsed? He therefore found comfort in the notion that modern European society was not as vulnerable to barbarian invasions as the Romans had been, because the invention of gunpowder made the military an industry of which the 'opulent and civilised nations' had a monopoly. Smith realised that the quality of military force changed with technological development; thus, the modern military, with its guns, was qualitatively different from the armies of the longbow and the pike. In Smith's writings, however, we also find the first inkling of the modern notion that you can pay for security by outspending your enemies. The 'opulent nations' can pay the insurance premium on defence and rest assured that history will not catch up with them. This entire line of argument is based on the premise that military force is a given quality that must be applied to a specific set of circumstances. This is why the insurance metaphor works so well. From the insurance perspective, change has become an argument

---

[46] Adam Smith, *An Inquiry into the Nature and Causes of the Wealth of Nations*, (Indianapolis, IN: Liberty Classics, 1981), 708.

for larger, more technologically adept forces in order to be ready for any eventuality. It is this confidence that 'whatever happens, we've got the Maxim gun and they don't', which the coinistas question. They believe the circumstances for security are profoundly changing, meaning that it is implausible that one can in fact have a capability for most eventualities, and it renders it impractical to invest in large, expensive forces that might do little by manoeuvres waiting for a conventional confrontation that never arrives. The effect of this conventional approach is to crowd out resources that could otherwise be spent on current operations or research and development in future technologies. This is all the more troubling if it is in fact true that the NBIC technologies will profoundly transform military technology. In that case, a qualitative change in technology equal to the one identified by Adam Smith, with the invention of guns and standing armies, will take place. In sum, neither the COIN perspective nor the technological perspective supports the notion that large conventional forces, organised by the modern military system, will create a sufficient buffer to ensure future security.

Future NBIC technologies might overcome this entire issue. The DARPA survey encourages one to regard robots as a system that would work together with human operators, thus providing a platoon or a squadron with a much larger radius and firepower than otherwise would be the case. What works as a force multiplier on the platoon level might also work for the force structure, as such. If automated systems can be stored for use in case of hostility, then the armed forces will no longer need to pay a huge premium for preparing for operations that in all likelihood will never happen. Furthermore, robots can function as reserves and be taken out of storage when small, agile units get into trouble. This buffer function will dramatically reduce costs, but one might speculate that it would also transform the military system profoundly. The need to prepare for a modern battle has transformed modern militaries into logistical organisations and training establishments. Autonomisation could dramatically reduce the need for such peacetime preparations and create a buffer to be used in time of war. This should reduce the tail significantly.

Even if the robots do not have to do push-ups, their human companions still need to train with the interface and, more importantly, commanding officers and their political masters must focus on how to use the force. The officers' attention would thus not be on the next exercise, but on how to use the dormant capabilities and towards which ends. This may result in a shift in focus from operations to strategy. This strategic focus would come to much use in the day-to-day operations of the military in ways which it does not today. Thus, it is entirely possible that automatisation will mean that the armed forces' day-to-day focus will resemble the COIN paradigm much more than the AirSea Battle paradigm. The technological buffer will enable the armed forces to focus on what is now occasionally rather condescendingly referred to as 'operations other than war'; that is, the entire spectrum of operations that allows armed forces to provide the final piece of the puzzle that enables an operation to be successful.

The modern military system defined the use of military force in a closed system in which one army was pitched against another. When this type of engagement is valued above any other, then the military's ability to engage in other types of conflicts is naturally less. What is perhaps less appreciated is how the modern system has disengaged the armed forces from other parts of national policy. That is not to say that within the current system, army engineers cannot be used for helping out after natural disasters or that an Air Force C130 cannot fly supplies to refugees. Even in these instances of the military doing 'civilian' jobs, however, these jobs are regarded as exceptions; operations which are carried out on the side. If the armed forces were able to operate outside the confines of the modern system, then operations involving close cooperation with civilian counterparts could more easily become the defining operations. Technology allows a more modular approach, which would see military force as a component in a government effort rather than an effort unto itself. The defining feature of such an effort would be to contribute to what Richard Thaler and Cass Sunstein term 'choice architecture'.[47] Within the current thinking on how to use new platforms like drones, the closed-system view on military strategy is used to build 'Skinner Boxes'. Boxing an opponent in using various punishments appeals to those schooled in the modern military system, but Thaler and Sunstein offer an alternative perspective on how to create contexts that encourage the realisation of new sets of values. Where the individual was a statistical category to Skinner, who was concerned with the environment that shaped individual choice, to Thaler and Sunstein the individual is the focus of the 'choice architect', who is developing techniques for making individuals aware of their choices. Thaler and Sunstein thus focus on how to make people aware of automated, unreflected ways of acting (they term it 'automatic systems') in order to allow them to choose otherwise. They term this approach 'nudging'. Nudging can create a new choice architecture in the organisation of the armed forces as well as in their operations.

In terms of operations, nudging is not as alien as it might initially sound. For example, deterrence is essentially a nudging strategy compelling an opponent to refrain from aggressive behaviour.[48] A nudging strategy will depend heavily on information that makes choice mapping possible. As opposed to the Skinner Box, which regulates behaviour and punishes individual transgressions (US drone strikes in Yemen and Pakistan are a case in point), the use of military force for nudging will be focused on creating a choice architecture that allows for terrorists to be neutralised without alienating the civilian population. The operations against Somali pirates offer an example of such a strategy, where a strong naval presence and the protection of individual ships by armed guards has dramatically reduced the incentives for pirates to board ships, while various on-land development projects have encouraged the pirates to opt out

---

[47] Richard Thaler and Cass Sunstein, *Nudge* (London: Penguin, 2009), 3.
[48] Lawrence Freedman, *Deterrence* (Cambridge: Polity Press, 2004).

of the business altogether. In the fight against piracy in the Gulf of Aden, military force was thus only one element amongst many in a coherent effort.[49]

Similarly, NATO strategy on Russia in the aftermath of the Ukraine crisis in 2014 needed to integrate conventional deterrence with economic sanctions and other measures in order to create a 'choice architecture' that would discourage the Kremlin from interventions in Ukraine and elsewhere. The Russian use of special forces, intelligence and agents among allied groups in Ukraine created the possibility for infiltrating below the horizon of the NATO's conventional deterrence. John Schindler of the Naval War College refers to this as 'special war'.[50] NATO thus needed to be able to deter special war as well as conventional war. While the Alliance was able to activate its military deterrence fairly quickly by deploying US Army units to the Baltic countries and Poland as well as air and naval assets, the ability to coordinate economic, social, political and military assets has proved more complicated. Since NATO had emphasised the need for a 'comprehensive approach' in operations in Afghanistan, it is perhaps surprising that such a comprehensive approach is so difficult to implement in the Alliance's back yard. Nonetheless, the notion of 'choice architecture' captures the demands on the Alliance much better than traditional notions of deterrence and thus emphasises the need for integrating military and civilian capabilities in a coherent security policy. As President Obama noted, 'both the Europeans and the United States have been consistent in calibrating sanctions that could provide a deterrent for the Russians, providing support to the Ukrainians, leaving open a path for resolving this problem diplomatically'.[51] In other

---

[49] Lars Bangert Struwe, 'Private Security Companies (PSCs) as a Piracy Countermeasure', *Studies in Conflict and Terrorism* 35 (2012), 588–96.

[50] John R. Schindler, 'How to Win Cold War 2.0', *Politico*, 25 March 2014, www.politico.com/magazine/story/2014/03/new-cold-war-russia-104954.html#.U1319cYbZ8D (28 April 2014).

[51] Barack Obama, *Press Conference with President Obama and President Park of the Republic of Korea*, Blue House, Seoul, Republic of Korea, 25 April 2014, The White House, www.whitehouse.gov/the-press-office/2014/04/25/press-conference-president-obama-and-president-park-republic-korea (28 April 2014).

words, the US government wanted to create a choice architecture that punished the Russian government and Russian elites for acting in certain ways by means of sanction, while engaging Russia diplomatically and otherwise at the same time. This was probably the most important reason why President Obama continuously argued that the conflict with Russia was no Cold War 2.0. Instead of the Skinner Box deterrence of the Cold War, the Obama administration and its European allies had to come up with a strategy that suited a globalised world where Russia was integrated in the world economy. Even if the United States and its allies seemed a little ruffled by the confrontation with Russia, they gradually and cautiously experimented with a strategy for dealing with Russia which was clearly informed by the ideas behind nudging.

The armed forces are distributing resources, procuring platforms and organising operations in ways that are, if not unrelated to what is happening in the world outside the military system, then over-determined by the system's needs. Treasuries attempt to control this process by reducing budgets in the hope that this will increase effectiveness; but when the system is the cause, the best the military can do in times of austerity is to do more with less – which only brings nearer the day described by Augustine, where the entire US armed forces share one plane. In order to break this logic, the armed forces need to be 'nudged' in another direction. The choices facing planners have to be posed in ways that allow for value creation instead of reliable but increasingly suboptimal outcomes. As mentioned, NBIC technologies may create a buffer that would make it possible to reassign resources to current, urgent operations, but this will only happen if the mindset of military planners is framed by something other than the modern military system.

Since nudging is about people realising what they really need, then one might imagine something like the Warfighter Ranking Scheme of Procurement. Instead of putting procurement in the purview of staff officers fighting their bureaucratic battles, the outcome of which has been a frightening regularity in which services get what, one could leave it to front-line personnel to make procurement priorities. They should be presented with future procurement options that would benefit not them, but their successors 10–15 years down the line and then prioritise procurement items based on their experiences in the field. Prioritisation would include the budgetary constraints in a process that might otherwise be a free-for-all. The Defence Minister might have good reasons to decide otherwise, but these decisions would put operational requirements centre stage. So would HR policies that nudged talented officers into front-line careers, rewarded cooperation with civilian agencies (perhaps even required work experience in, for example, an NGO or a civilian accountancy firm) and in other ways ensured that officers regard themselves as part of an open, post-modern, military system, where their capabilities can be used in different settings. HR policies prioritising operational experience and practical results will work against the current systemic logic by trumping actual achievements. To further that aim, what is actually achievable and achieved by military forces needs to be much in focus. Civilian counterparts and people on the ground should be asked to evaluate the officers themselves as well as the operational results. Preferably, one should ask the enemy. How better to evaluate the effort in, say, Afghanistan's Helmand Province, than to ask the Taliban about their evaluation of particular units or operational strategy? Such questions are obviously an integrated part of intelligence gathering, but the purpose is usually to get an impression of the enemy's capabilities rather than one's own capability. Theo Farrell and Antonio Giustozzi put such questions to the Taliban, and the Taliban's assessment is very instructive. For example, the Taliban's assessment of the British attempt to eradicate poppy fields in Helmand was that it produced much resentment and fresh recruits for the insurgency: 'When they started destroying the opium fields, the people – landowners, farmers, poor people – everyone became angry. And they started fighting.'[52]

---

[52] Theo Farrell and Antonio Giustozzi, 'The Taliban at War: Inside the Helmand Insurgency, 2004–2012', *International Affairs* 89 (2013), 852.

Such operational mistakes must have consequences if military forces and individual officers are to learn from them. Thomas Ricks has argued that relieving general officers for operational shortcomings would strength the overall performance of the US Army.[53] This is nudging with a large stick. But as Ricks notes, it can only be one part of a system of incentives that make operational experience matter for how the armed forces are run. Thus, military nudging would need to strengthen feedback loops and thus make the armed forces more responsive to outside demands. The aforementioned 'nudges' are only examples of how one can create military organisations that are more innovative and an officer corps that is more curious.

Use the transformation of the military from a service to a profession of arms as an opportunity to transcend the warrior–soldier dichotomy

Nudging the armed forces to be less self-referential and more focused on actual operational experience should serve to make the military leadership better at prioritising resources. In times of scarcity, making priorities becomes more important. Designing the armed forces as part of the entire government's security capacity is a way to make priorities that place the armed forces' budget in context and thus focus on what the armed forces are supposed to achieve rather than on how the armed forces can spend the taxpayers' money. In this context, the most important question is not what the military system needs in order to function, but rather how the military system functions in synergy with other government departments. In business model terms, it is about identifying what the customer wants. Unfortunately, the average staff officer is not well prepared for thinking in terms of demand and being part of a larger 'strategic community'. The shopping-list approach to strategy is a reflection of how much military strategising is focused on what is needed for the modern system to function

---

[53] Thomas E. Ricks, *The Generals: American Military Command from World War II to Today* (New York: Penguin, 2012), 449–50.

reliably rather than on creating value for those who actually make the decisions and pay the bills. Christopher Paparone and George Reed describe this as the 'competency trap'. Having been taught the ways of the system since they were officer cadets, the average staff officer knows the military profession so well that they are often unqualified to learn to operate in new ways.[54] The closed nature of the military system creates equally closed education and career paths that give individual officers high competences in operating the system but few competences in terms of appreciating the purposes the system serves. This competency trap thus manifests itself as a lack of self-awareness and ability to compare one's competences with those of civilian society. In a study of US officers, Ender finds, as mentioned in Chapter 2, that 77 per cent of military leaders believe that 'civilian society would be better off if it adopted more of the military's values and customs'.[55] There is much to respect in the values and customs of military officers, but given the armed forces' inability – in the United States and the rest of the Western world – to innovate and make ends meet, it would probably be nothing less than an economic catastrophe if the rest of the public sector, not to mention the private sector, was to be run like the armed forces.

In order to escape the 'competency trap', military personnel, especially officers, must be trained in innovation and rewarded for producing value rather than reliable results. As mentioned above, there are ways for the armed forces to nudge human resources in that direction. This is also a matter of soldier identity. This means defining soldiering as a profession rather than a service. Defined in terms of a service, the military is a tightly knit band of brothers defined by the risks servicemen and women take in the service of their country. It is this service for ideas of the nation or ideology

[54] Christopher R. Paparone and George Reed, 'The Reflective Military Practitioner: How Military Professionals Think in Action', *Military Review* 88 (2) March–April 2008, http://usacac.army.mil/CAC2/MilitaryReview/Archives/English/MilitaryReview_20080430_art011.pdf (7 June 2012), 69.

[55] Morten G. Ender, *American Soldiers in Iraq: McSoldiers or Innovative Professionals?* (New York: Routledge, 2009), 50.

that confers meaning to the practice of soldiering and justifies the risks and the sacrifices involved. From a professional perspective, the order of meaning is reversed. The professional soldier derives identity from being an expert in the technologies and techniques of soldiering. This opens the profession to anyone able to acquire the soldier's skills, which is clearly one reason why professional armed forces are less homogeneous than the band of brothers. Such forces have room for women, homosexuals and ethnic minorities, because it is skill rather than character that defines the professional military community. Anthony King observes this development with regard to the European armed forces. King argues that the mass conscripted armies which used to characterise the European militaries were homogeneous and had a strong identity. This possibly made the individual soldiers less effective.[56] The new armies, honed by international operation in, for example, Afghanistan, are less homogeneous, as they reflect societies that are more diverse. These forces thus include ethnic minorities and more women. 'Europe's armed forces – which are increasingly all-volunteer professional – may be increasingly integrated not by concepts of national identity, civic duty or personal bonds, but rather by a concrete professional ethos', King argues.[57] This professionalisation is much more pronounced in the United States. Chairman Dempsey talks about 'a profession of arms', arguing that the American servicemen and women constitute 'a profession of experts in the use of military power to defend America'.[58] The chairman actually wants it both ways – in his view, the American soldier is a paragon of republican virtue as well as a professional expert.

Perhaps it would be more accurate to describe the political and patriotic motives for joining the armed forces as something distinct from actually working there. The political and national context is an

---

[56] Anthony King, *The Transformation of Europe's Armed Forces: From the Rhine to Afghanistan* (Cambridge University Press, 2011), 227–30.

[57] *Ibid.*, 235.

[58] US Joint Chiefs of Staff, *Chairman's Direction*, 9.

important framing of the situation that defines the military practice, but the military practice is itself constituted by what Jean Lave and Etienne Wenger describes as a 'generative social practice'.[59] Lave and Wenger describe practices that generate learning by creating an environment where 'understanding and experience are in constant interaction – indeed, are mutually constitutive'.[60] Such 'situated learning' occurs in a carpenter's workshop, for example. In the course of an apprenticeship, one is not only training certain skills, one becomes part of a 'community of practice' which defines not only *what one does* but also *who one is* in terms of a professional identity. 'Learning is never simply a process of transfer or assimilation', Lave and Wenger note.[61] The apprentice engages with the master and the journeymen in learning. The apprentice not only learns his trade, but learns it by working alongside the journeymen. In that learning process, the journeymen renegotiate the nature of the trade itself, confirming its skills and codes, as well as creating new ones, together with the apprentice. A professional military increasingly comes to see the profession of arms as a 'community of practice'. Defined in this manner, the military procession is a learning process wherein experience and innovation are mutually constitutive. Focusing on learning the military trade in a community rather than as an identity which the soldier assimilates in boot camp, the military identity does not have to be regarded as fixed or reified. Thus, from this perspective, there is no conflict between being a warrior in the Nietzschean sense and operating the new NBIC technologies. The power of the automated arsenal will undoubtedly inspire the human soldiers to strike 'Nietzschean poses' in order to find meaning in their service alongside more physically capable robots. The special forces soldiers, with their highly specialised training, strong ethos and specific missions, are much more the shape of things to come than the UAV operators, who are perhaps best compared to crossbowmen – highly skilled operators of

---

[59] Jean Lave and Etienne Wenger, *Situated Learning: Legitimate Peripheral Participation* (Cambridge University Press, 1991), 35.
[60] *Ibid.*, 51–2. [61] *Ibid.*, 57.

a weapons system which is decisive for a time but which technological development quickly passes by. New technologies are part of the constitutive process of defining and redefining what it means to be a soldier. Thus, there is no loss of identity when adopting new platforms or procedures, because being able to adapt and innovate is a part of the military identity if constituted as a community of practice. Framing military identity in those terms is thus an important element in the creation of a value-focused military.

Utilise global networks

The modern military system conserves the armed forces in a balance between services and functions by chain-linking issues in ways that make reform seem almost impossible. The fact that Western military systems are closely linked by joint doctrines in NATO, educational practices and linked systems of procurement reinforces the impression of an impregnable, unreformable system. This is perhaps especially so when observed from a small or medium-sized European country that basically has to take the standards of the military system as they are. Yet the fact that issues and national armed forces are chain-linked can be turned into an advantage for those calling to design a military system for the twenty-first century. Scott Anthony points out that while large organisations have traditionally had problems with thinking in terms of validity and embracing innovation, which therefore had to take place in small start-ups housed in daddy's garage, innovation today often takes place in 'the new corporate garage' of large companies that have found ways to put their superior resources at the disposal of 'corporate catalysts'.[62] From this perspective, small or medium-size militaries can be regarded as agents of change rather than 'bonsai armies'. NATO is in fact working on the creation of a 'corporate garage' mentality via the 'Smart Defence Initiative'. In itself, this initiative may be little more than a slogan,

---

[62] Scott D. Anthony, 'The New Corporate Garage', *Harvard Business Review*, September 2012, 45–53.

but it is exactly this type of slogan that can free creativity and give 'corporate catalysts' the traction they need to innovate. The NATO heads of state and government conceded as much at their Chicago Summit in 2012, where their communiqué first listed a number of multilateral initiatives and then went on to argue that 'Smart Defence is more than this. It represents a changed outlook, the opportunity for a renewed culture of cooperation in which multinational collaboration is given new prominence as an effective and efficient option for developing critical capabilities'.[63]

The Smart Defence Initiative is one way to use the fact that the Western military system is so integrated. In its relation to allies, the United States has seemed more interested in offloading assignments in order for the Department of Defense to focus on engaging in real warfare, not just helping others to defend themselves but also helping others to defend the United States,[64] than on actually using allies as a force-multiplier. One exception to this approach seems to be how the US Special Forces Command is creating a global network of allied special operations forces.[65] It is instructive how the forces which are in the highest demand are organising themselves in global networks in order to fulfil that demand. As Anthony King points out, however, this has been the case with European armed forces since the end of the Cold War. The Danish armed forces offer a case in point. The Danish Army has deployed a battle group of two companies of armoured infantry supported by a contingent of Leopard II tanks to the British Task Force Helmand. Combat engineers, fire support, ordnance disposal and a civil–military relations team have supported these forces with the occasional deployment of special forces teams

[63] NATO, *Summit Declaration on Defence Capabilities: Toward NATO Forces 2020*, 20 May 2012, www.nato.int/cps/en/natolive/official_texts_87594.htm?mode=pressrelease (12 December 2012), §8.

[64] Robert Gates, 'Helping Others Defend Themselves: The Future of U.S. Security Assistance', *Foreign Affairs*, May/June 2010, 2–6.

[65] 'Socom Officials Work on Plan for Global Network', American Forces Press Service, 2013, www.defense.gov/news/newsarticle.aspx?id=120193 (11 December, 2013).

in support. A Danish medical team has also supplemented the British effort from time to time. This small force has provided a remarkable punch. In 2011, the teeth-to-tail ratio for the Danish battle group in Helmand was 2:1, with 65 per cent of the Danish forces in teeth functions and 35 per cent in the tail.[66] Compared to the US numbers, this is a remarkably high ratio of combat troops. From the Danish perspective, this was a demonstration of how a continuous focus on costs had enabled the Danish armed forces to increase quality for less money. This was also true in terms of the costs of the operation. In 2009, Minister of Defence Gitte Lillelund Bech told Parliament that the annual cost of the Afghanistan operation was DKK 2 billion,[67] meaning that the 750 soldiers in Afghanistan that year had an individual operating cost of DKK 2.6 million (USD 452,000). That is cheap compared to the estimated price of USD 1 million for an American soldier in Afghanistan. Deploying a battalion means that the Danish Army does not have to provide for the staff, support and logistics on the brigade or division levels. In effect, the Danes provided the teeth and let others pay for the tail. Since the British taxpayer had already paid for the tail of Task Force Helmand, accommodating Danish and other allied units might actually make good economic sense, since it makes the sunk costs of establishing the operation count for more. Limited capabilities have made networking necessary. King observes that the same international networks that sociologists have identified with the changing industrial relations in globalisation increasingly structure European armed forces.[68] Manuel Castells notes how these networks are the unit of analysis rather than the individual

---

[66] The numbers are based on Danish Army Staff calculations, which break down as follows: DANCON ISAF 11 (January–July 2011), 693 troops in total, 456 in teeth functions, including two companies of infantry, tanks, engineers, fire support, ordnance disposal and a civil–military relations team, and 237 in tail functions, including staff, logistics and military policy. DANCON ISAF 12 (July 2011–February 2012) was 699 troops in total, with 458 in teeth functions and 241 in the tail.

[67] *Folketinger S1899, Samling 2009–10: Om udgifter til missionen i Afghanistan.*

[68] King, *The Transformation of Europe's Armed Forces*, 97.

firms making up the networks.[69] If this perspective is to be put to use in the development of Western armed forces, then the integrated nature of the modern military system can be used as an asset in its transformation.

Perhaps these networks are described too narrowly in terms of the military cooperation in NATO and other US-managed global defence systems. Tarak Barkawi and Shane Brighton argue that current British grand strategy fails to utilise the fact that Britain is the hub of a post-colonial network, because British elites do not imagine Britain on a global scale. According to Barkawi and Brighton, British strategy is informed by a reading of history that describes national development in isolation from international events, which makes the British seem even more valiant when international events catch up with them and they have to stand up to Napoleon and Hitler on their own.[70] As David Edgerton points out, this notion that the United Kingdom fought the Second World War alone from June 1940, when France fell, until June 1941, when the United States entered the war, is a misrepresentation of the facts. In fact, the British Empire provided a massive resource base, which might be alone against Germany but which nevertheless represented the greatest power in the world. British power was no more limited to the resource base of the British Isles than was the British Army dependent on the forces evacuated from Dunkirk. The 1940 plan for expanding the Army to fifty-five divisions only expected thirty-four of these to be recruited in Britain. Nine divisions were to be raised in India, four in the African colonies, three in Canada and Australia, respectively, and South Africa and New Zealand would provide one division each. The Indian Army would end on a strength of 2.5 million soldiers. These troops fought in all theatres of war.[71] Later, this networked strength was largely

<hr>

[69] Manuel Castells, *The Rise of the Network Society* (London: Blackwell, 2000), 501.

[70] Tarak Barkawi and Shane Brighton, 'Brown Britain: Post-Colonial Politics and Grand Strategy', *International Affairs* 89 (2013), 1115–19.

[71] David Edgerton, *Britain's War Machine: Weapons, Resources and Experts in the Second World War* (Oxford University Press, 2011), 47–85.

ignored in a narrative that focused on British weakness, isolation and eventual decline rather on strength, networks and growth in absolute terms, even if not in relative terms. Edgerton terms this narrative 'declinist histories',[72] and Barkawi and Brighton point out that this narrative of decline has been reinvigorated by the defeats in Iraq and Afghanistan and the economic decline of the West relative to the East.[73] This perspective is as wrong about the future as it is about the past, Barkawi and Brighton argue, chiding British strategists for seeing decline instead of opportunity in current developments – 'their conservatism and realism work against them and blind them to actually existing social and political trends, processes and potentialities'.[74] In other words, they focus on reliability rather than on value creation. Barkawi and Brighton point out the values in contemporary British society. British society is embedded in international networks and British citizens in interrelationships across the globe. These networks are the product of empire, but they are also a future reality in which Britain is more ethnically diverse and plugged in to more communities than before, and 10 per cent of Britons live abroad, making their contribution to other communities, which in turn are becoming more ethnically diverse and plugged in to the global network. This global interconnectedness is the proper context for national power, Barkawi and Brighton argue, and they point out how interconnectedness can be used as a resource for national power.[75]

Even if British imperial history is unique, the United Kingdom is hardly the only European nation living in post-colonial times, nor are the former imperial powers the only European powers which have had their demography and overseas connections radically reconfigured. Large movements of people and reconfigurations of identity have marked the post-Cold War years in Europe. As mentioned above, this is reflected in the make-up of the armed forces and is an important impetus to adopting a professional approach to

---

[72] *Ibid.*, 6.   [73] Barkawi and Brighton, 'Brown Britain', 1109.
[74] *Ibid.*, 1116.   [75] *Ibid.*, 1122–3.

soldiering. The national nature of the armed forces and the strategies they implement have been largely taken for granted, however. For small to medium-sized European powers with little or no colonial legacy, this presents a particular challenge. Nonetheless, these states are those with the greatest interest in making networks force multipliers. On the strategic level, the European powers are fundamentally challenged by the fact that armed force is not the most relevant means to provide security. While the fact that the prospect of military confrontation is as distant as it has ever been on the European continent is obviously a good thing for those living there, it renders the question of the European militaries' business all the more acute. What problem is the European armed forces solving? Unable to answer that question to the satisfaction of the European treasuries, the budgets of the European armed forces are being cut across the board. For small to medium-sized European powers, these cuts have mainly been on the procurement title of the budget,[76] which may in time challenge the ability of these countries to adopt NBIC technologies. This will not only render them technologically obsolete, it will also diminish their relevance to the US armed forces, which are 'pivoting' to Asia in the realisation that future confrontations will take place there rather than in Europe. The Europeans are increasingly feeling that they are home alone, meaning that the innovation of the armed forces, which has been driven by the Americans through NATO, will not necessarily take place to the extent known previously.

Barkawi and Brighton's approach to strategy makes clear that the Europeans are in fact not home alone, but rather that the nature of that home, as well as the neighbourhoods it connects to, has been radically transformed; a fact which is largely ignored, even in otherwise well-informed and reflected comments on European foreign policy, which take the notion of Europe, European security and

---

[76] Mikkel Vedby Rasmussen and Jon Rhabek-Clemmensen, *Trends in NATO Defence Budgets*, mimeo, (Copenhagen: Centre for Military Studies, 2014).

European values for granted.[77] The inability to formulate common European positions is widely regarded as a result of differences in national strategic cultures;[78] thus, cultural heterogeneity is a problem which must be overcome rather than a potential resource which should be used productively.[79] One reason for this view is the way the European Union has defined integration as Europeanisation. To some extent, European integration has privileged connections between Europeans over global connections, but anyone walking through any major European city can see for themselves that these connections are there. Today, these connections are unseen and unmentioned in strategic discourse; if they were seen and mentioned, then it would be obvious that neither European interests nor European resources were confined to Europe itself. In fact, the notion of what constituted Europe as well as individual European states would be profoundly reconfigured in terms of global connectedness and interrelational identities. If national security were defined in those terms, armed force would much more readily be seen as a part of a national security capacity; a capacity which in itself was defined by cooperation with other national militaries. In the absence of a nationalistic and conservative reading of the function of military forces, it would be much easier to regard multinational military cooperation in terms of function rather than a sovereignty issue. Moreover, cooperation with non-Western militaries could be redefined from the paternal relationship, as in capacity building, to partnerships dealing with common problems. This approach would move the logic of cooperation with non-Western forces from a problem focus ('their militaries do not work and we have to make them work') to an issue focus ('how do

[77] For one example, see Susi Dennison *et al.*, *Why Europe Needs a New Global Strategy*, European Council of Foreign Relations, Policy Brief, October 2013, http://ecfr.eu/page/-/ECFR90_STRATEGY_BRIEF_AW.pdf (4 December 2013).

[78] Olivier de France and Nick Witney, *Europe's Strategic Cacophony*, European Council of Foreign Relations, Policy Brief, April 2013, http://ecfr.eu/page/-/ECFR77_SECURITY_BRIEF_AW.pdf (4 December 2013).

[79] Jürgen Habermas, 'The European Nation State: Its Achievements and Its Limitations. On the Past and Future of Sovereignty and Citizenship', *Ratio Juris 9* (1996), 125–37.

we combat al-Qaeda in North Africa?'). Such an approach would go a long way towards making national security establishments and their militaries value-producing.

Respect the paradox of military power

The potent power of armed force is deceptive. The paradox of military power is that no other government instrument is as decisive but the nature of armed force is such that the decisiveness is as partial as it is ephemeral. If the Western armed forces successfully transform themselves to more value-producing institutions, it will become even more important to recognise the paradox of military power. A defence establishment in close cooperation with its political customers, with other government departments, civilian actors and defence contractors will be able to promise a dazzling array of new technologies and strategies. In producing these platforms and ideas as well as evaluating their merits, it will be important to remember that the paradox of military power also applies to nano-weapons and whole-of-government approaches. In fact, the realisation that military decisiveness is partial and ephemeral is arguably a precondition for producing strategies that are more than promises but in fact deliver security.

As cautious as one will need to be if the armed forces are actually able to reform themselves and escape the law of diminishing returns which is governing them now, there will be even more cause for concern if reform is not forthcoming. The modern system was born on the battlefield after the ineffective strategies of a bygone era had sent thousands upon thousands to their death. These individual tragedies were only compounded by the disgust of the military as an institution that could organise such slaughter.[80] This meant that the armed forces lost the respect and understanding of civilian society. In many ways, a more reflective approach to the use of armed force was

---

[80] James Sheehan, *The Monopoly of Violence: Why Europeans Hate Going to War* (London: Faber and Faber, 2008).

a good thing, but the lack of understanding of the nature of modern war also resulted in a carefree attitude to military power that led to the armed forces being neglected in the West and to militarism in Germany. This was a combination that led the world to disaster for the second time. Today, the case for military reform is based on the premise that neither demographic instability and globalisation nor the advent of NBIC technologies should lead to disaster. This will largely depend on the ability to design strategies for the armed forces that enable them to provide security together with other parts of the national government and in international networks of allies. This has to be done before today's security concerns becomes tomorrow's crises. Otherwise, we risk the post-modern military system being born in carnage, just as the modern system was.

Bibliography

ABC News, 'This Week' Transcript: Defense Secretary Leon Panetta, 2012: 2, http://abcnews.go.com/Politics/week-transcript-defense-secretary-l eon-panetta/story?id=16415968&page=2#.T9BMB451J0s (7 June 2012).

Alexander, Michael and Timothy Garden, 'The Arithmetic of Defence Policy', *International Affairs* 77 (2001), 509–29.

American Forces Press Service, 'Socom Officials Work on Plan for Global Network', www.defense.gov/news/newsarticle.aspx?id=120193 (11 December, 2013).

Andrzejewski, Stanislaw, *Military Organization and Society*, 2nd edition (London: Routledge & Kegan Paul, 1968).

Anthony, Scott D., 'The New Corporate Garage', *Harvard Business Review*, September 2012, 45–53.

Army Force Management School, *Department of Defense Planning, Programming, Budgeting, And Execution (PPBE) Process/Army Planning, Programming, Budgeting, And Execution (PPBE) Process*, 2006, www.acqnotes.com/ Attachments/Army%20PPBE%20Executive%20Primer.pdf (18 October 2014).

Army Times, 'The PowerPoint Rant That Got a Colonel Fired', 2 September 2010.

Ashworth, G. J., *War and the City* (London: Routledge, 1991).

Augustine, Norman R., *Augustine's Laws* (New York: Viking, 1986).

The Aviationist, 'F-22 Raptor Kill Markings Shown Off by German Eurofighter Typhoons', http://theaviationist.com/2012/07/23/f-22-raptor-kill-markings/ (13 September 2012).

Barkawi, Tarak and Shane Brighton, 'Brown Britain: Post-Colonial Politics and Grand Strategy', *International Affairs* 89 (2013), 1109–23.

Barno, David, Nora Bensahel, Matthew Irvine and Travis Sharp, *Sustainable Pre-eminence: Reforming the US Military at a Time of Strategic Change* (Washington, DC: Center for a New American Century, 2012).

BBC News, 'Text of Bush's Act of War Statement', 12 September 2001, http:// news.bbc.co.uk/2/hi/americas/1540544.stm (19 November 2013).

Berlin, Isaiah, 'Two Concepts of Liberty', in *The Proper Study of Mankind: An Anthology of Essays* (London: Pimlico, 1998).

Biddle, Stephen, *Military Power: Explaining Victory and Defeat in Modern Battle* (Princeton University Press, 2004).

Bolman, Lee G. and Terrence E. Deal, *Reframing Organizations: Artistry, Choice and Leadership*, 3rd edition (San Francisco, CA: Jossey-Bass, 2003).

Boot, Max, *The Savage Wars of Peace: Small Wars and the Rise of American Power* (New York: Basic Books, 2003).

Bowden, Mark, *The Finish: The Killing of Osama Bin Laden* (New York: Grove Press, 2013).

Bowden, Mark, 'The Killing Machines', *The Atlantic*, September 2013.

Bower, Joseph L. and Clayton M. Christensen, 'Disruptive Technologies: Catching the Wave', *Harvard Business Review*, January–February (1995), 43–53.

Bradley, Omar, 'Leadership', *Parameters* (Winter 2010), www.carlisle.army.mil/usawc/Parameters/Articles/2010winter/Bradley.pdf (7 June 2012).

Breitenbauch, Henrik Ø., *Kompas og Kontrakt* (Copenhagen: Danish Institute for Military Studies, 2008).

British Army, *Modernising to Face an Unpredictable Future: Transforming the British Army*, July 2012, www.army.mod.uk/documents/general/Army2020_brochure.pdf (27 September 2012).

Brodesser-Akner, Taffy, 'Max Brooks Is Not Kidding About the Zombie Apocalypse', *New York Times*, 21 June 2013.

Brooks, Max, *World War Z: An Oral History of the Zombie War* (New York: Three Rivers Press, 2006).

Brooks, Rosa, 'Obama Needs a Grand Strategy', *Foreign Policy*, 2012, www.foreignpolicy.com/articles/2012/01/23/obama_needs_a_grand_strategy (24 January 2012).

Bumiller, Elisabeth, 'We Have Met the Enemy and He Is PowerPoint', *New York Times*, 26 April 2010, www.nytimes.com/2010/04/27/world/27powerpoint.html (18 October 2012).

Bumiller, Elisabeth, 'West Point is Divided on a War Doctrine's Fate', *New York Times*, 27 May 2012, www.nytimes.com/2012/05/28/world/at-west-point-asking-if-a-war-doctrine-was-worth-it.html (7 June 2012).

Bungay, Stephen, *Alamein* (London: Aurum Press, 2002).

Bungay, Stephen, *The Art of Action: How Leaders Close the Gaps Between Plans, Actions and Results* (London: Nicholas Brealey Publishing, 2011).

Burton, J. B., 'The Officer Critical Skills Retention Bonus', *Small Wars Journal*, 18 July 2007, http://smallwarsjournal.com/blog/the-officer-critical-skills-retention-bonus (24 January 2012).

Carter, Aston B. 'Running the Pentagon Right: How to Get the Troops What They Need', *Foreign Affairs*, January/February 2014.

Castells, Manuel, *The Rise of the Network Society* (London: Blackwell, 2000).

Center for Disease Control, www.cdc.gov/phpr/zombies.htm (13 November 2013).

Chandrasekaran, Rajiv, *Little America: The War Within the War for Afghanistan* (London: Bloomsbury, 2012).

Clark, I. F., *Voices Prophesying War 1763–1984* (Oxford University Press, 1966).

Clausewitz, Carl von, *On War*, edited and translated by Michael Howard and Peter Paret, (Princeton University Press, 1976 [1832–34]).

CNN.com, 'Brazilian Army Occupies Rio Shantytown Ahead of World Cup', 24 April 2014, http://edition.cnn.com/2014/04/24/sport/football/brazil-world-cup-favela-slums/ (28 April 2014).

Cohen, Eliot A., *Supreme Command: Soldiers, Statesmen, and Leadership in Wartime* (New York: The Free Press, 2003).

Coker, Christopher, *Barbarous Philosophers: Reflections on the Nature of War from Heraclitus to Heisenberg* (London: Hurst, 2010).

Coker, Christopher, *War and the 20th Century* (London: Brassey's, 1994).

Coker, Christopher, *The Warrior Ethos: Military Culture and the War on Terror* (London: Routledge, 2007).

Collier, Paul, *The Bottom Billion* (Oxford University Press, 2008).

Conetta, Carl, *An Undisciplined Defense: Understanding the $2 Trillion Surge in US Defense Spending*, Project on Defense Alternatives, PDA Briefing Report #20, 18 January 2010, www.comw.org/pda/fulltext/1001PDABR20.pdf (2 January 2013).

Congressional Budget Office, *Costs of Military Pay and Benefits in the Defense Budget* (Washington, DC: Congress of the United States, 2013).

Cornish, Paul, *Strategy in Austerity: The Security and Defence of the United Kingdom*, 16 October 2010, www.chathamhouse.org/publications/papers/view/109490 (9 September 2014).

Creveld, Martin van, *Fighting Power: German and US Army Performance 1939–1945* (Westport, CT: Greenwood Press, 1974).

Creveld, Martin van, *Technology and War: From 2000 BC to the Present* (New York: The Free Press, 1991).

Danish Parliament, *Folketinger S1899, Samling 2009–10: Om udgifter til missionen i Afghanistan.*

Davis, Mike, *Planet of Slums* (London: Verso, 2007).

Debord, Guy, *Society of the Spectacle* (Detroit: Black and Red, 1983).

Defense Business Board, *Managing DoD Under Sustained Topline Pressure*, FY10-2, January 2010, www.dtic.mil/cgi-bin/GetTRDoc?AD=ADA556844 (30 January 2014).

Delillo, Don, *Point Omega* (London: Picador, 2010).

Deloitte, *Energy Security: America's Best Defense. A Study of Increasing Dependence on Fossil Fuels in Wartime and its Contribution to Ever Higher Casualty Rates*, 2009, www.deloitte.com/assets/Dcom-UnitedStates/Local%20 Assets/Documents/AD/us_ad_EnergySecurity052010.pdf.

Dennison, Susi, Richard Gowan, Hans Kundnani, Mark Leonard and Nick Witney, *Why Europe Needs a New Global Strategy*, European Council of Foreign Relations, Policy Brief, October 2013, http://ecfr.eu/page/-/ECFR90_ STRATEGY_BRIEF_AW.pdf (4 December 2013).

Deptula, David A., *Effects-Based Operations: Change in the Nature of Warfare*, February 2001, Aerospace Education Foundation, www.airforce-magazine.com/ SiteCollectionDocuments/TheDocumentFile/Strategy%20and%20Concepts/ EBO_deptula_2001.pdf (6 January 2012).

Doz, Yves, and Mikko Kosonen, *Fast Strategy: How Strategic Agility Will Help You Stay Ahead of the Game* (Harlow: Wharton School Publishing, 2008).

The Economist, 'The Cost of Weapons: Defence Spending in a Time of Austerity', 26 August 2010, www.economist.com/node/16886851 (6 January 2012).

The Economist, 'Defence Spending: Always More, or Else', 1 December 2011, www.economist.com/blogs/democracyinamerica/2011/12/defence-spending (20 January 2012).

Edgerton, David, *Britain's War Machine: Weapons, Resources and Experts in the Second World War* (Oxford University Press, 2011).

Eikenberry, Karl, 'The Limits of Counterinsurgency Doctrine in Afghanistan', *Foreign Affairs*, September/October 2013.

Ender, Morten G., *American Soldiers in Iraq: McSoldiers or Innovative Professionals?* (New York: Routledge, 2009).

European Defence Agency, *National Breakdowns of European Defence Expenditure*, www.eda.europa.eu/Libraries/Documents/EDA_-_National_Defence_ Expenditure_in_2005.sflb.ashx (25 September 2012).

European Defence Agency, *National Defence Data 2012 of the EDA Participating Member States*, www.eda.europa.eu/docs/default-source/finance-documents/ national-defence-data-2012.pdf (25 September 2012).

Farrell, Theo and Antonio Giustozzi, 'The Taliban at War: Inside the Helmand Insurgency, 2004–2012', *International Affairs* 89 (2013), 845–71.

Farrell, Theo, Frans Osinga and James A. Russell (eds.) *Fighting the Afghanistan War* (Stanford University Press, 2013).

Finkelstein, Robert and James Albus, *Technology Assessment of Autonomous Intelligent Bipedal and Other Legged Robots*, Final Report submitted to Dr Alan Rudolph, Defence Sciences Office, Defence Advanced Research Projects

Agency, November 2004, www.robotictechnologyinc.com/images/upload/
file/Humanoid%20Robot%20Final%20Report%20Rev%2013%20May%20
03%20&%2021%20Nov%2004.pdf (28 August 2012).

FOX News, 'Panetta Warns of Smallest Air Force Ever if Deep Defense Cuts
Made', 20 January 2012.

France, Olivier de and Nick Witney, *Europe's Strategic Cacophony*, European
Council of Foreign Relations, Policy Brief, April 2013, http://ecfr.eu/page/-/
ECFR77_SECURITY_BRIEF_AW.pdf (4 December 2013).

Freedman, Lawrence, *Strategy: A History* (Oxford University Press, 2013).

Freedman, Lawrence, *Deterrence* (Cambridge: Polity Press, 2004).

Fuerth, Leon S., *Anticipatory Governance: Practical Upgrades*, forwardengage-
ment.org, 2011, www.forwardengagement.org (22 November 2012).

Gates, Robert M., 'A Balanced Strategy: Reprogramming the Pentagon for a New
Age', *Foreign Affairs*, January (2009).

Gates, Robert M., *Duty: Memoirs of a Secretary at War* (New York: Random
House, 2014).

Gates, Robert M., 'Helping Others Defend Themselves: The Future of U.S.
Security Assistance', *Foreign Affairs*, May/June 2010, 2–6.

Gates, Robert M., Speech at United States Military Academy (West Point, NY). As
Delivered by Secretary of Defense Robert M. Gates, West Point, NY, 25 February
2011. Defense.gov Speech, www.defense.gov/speeches/speech.aspx?speechid=1539
(24 January 2012).

Gentile, Gian P., 'Eating Soup with a Spoon', *Armed Forces Journal*, September
2007, www.armedforcesjournal.com/eating-soup-with-a-spoon/ (7 June 2012).

Gentile, Gian P., 'A Strategy of Tactics: Population-centric COIN and the Army',
Parameters: The US Army War College Quarterly, Autumn 2009, 5–17.

Google Ngram Viewer. 2012. Google Ngram Viewer, http://tiny.cc/laoclwhttp://
tiny.cc/laoclw (28 September 2012).

The Guardian, 'Afghan Schools and Clinics Built by British Military Forced
to Close', 27 September 2012, www.guardian.co.uk/world/2012/sep/27/
afghan-schools-clinics-built-british-close?intcmp=239 (6 November 2012).

The Guardian, 'Army Cuts: How Have UK Armed Forces Personnel Numbers
Changed Over Time?' 5 July 2012, www.guardian.co.uk/news/datablog/2011/
sep/01/military-service-personnel-total#data (25 September 2012).

The Guardian, 'Stanislav Andreski: Forthright Founder of Reading University
Sociology Department', 7 November 2007.

Habermas, Jürgen, 'The European Nation State: Its Achievements and Its
Limitations. On the Past and Future of Sovereignty and Citizenship', *Ratio
Juris* 9 (1996), 125–37.

Hagel, Chuck, *FY15 Budget Preview*, as Delivered by Secretary of Defense Chuck Hagel, Pentagon Press Briefing Room, 24 February 2014, www.defense.gov/Speeches/Speech.aspx?SpeechID=1831 (26 February, 2014).

Hagel, Chuck, *Statement on Strategic Choices and Management Review*, as Delivered by Secretary of Defense Chuck Hagel, Pentagon Press Briefing Room, 31 July 2013, www.defense.gov/speeches/speech.aspx?speechid=1798 (26 November 2013).

Hammes, T. X., 'Dumb-dumb Bullets', *Armed Forces Journal*, July 2009, www.armedforcesjournal.com/2009/07/4061641/ (10 May 2012).

Hammes, T. X., *The Sling and the Stone: On War in the 21st Century* (Minneapolis, MN: Zenith, 2006).

Harrison, Todd, *Estimating Funding for Afghanistan*, 1 December 2009, The Center for Strategic and Budgetary Assessments.

Hart, Peter, *1918: A Very British Victory* (London: Phoenix, 2009).

Hastings, Max, *Armageddon: The Battle for Germany 1944–45* (London: Pan Books, 2004).

Hastings, Max, *Bomber Command* (London: Pan, 2010).

Hastings, Max, *Finest Years: Churchill as Warlord* (London: HarperPress, 2009).

Hastings, Michael, 'The Runaway General', *Rolling Stone*, 2010, www.rollingstone.com/politics/news/the-runaway-general-20100622?page=2 (23 May 2012).

Henriksen, Rune, 'Warriors in Combat: What Makes People Actively Fight in Combat', *Journal of Strategic Studies* 30(2) (2007), 187–223.

Herwig, Holger H., 'The Battlefield Revolution, 1885–1914', in MacGregor Knox and Williamson Murray (eds.), *The Dynamics of Military Revolution 1300–2050* (Cambridge University Press, 2001), 114–31.

Herwig, Holger H., *The Marne – The Opening of World War I and the Battle That Changed the World* (New York: Presidio Press, 2012).

Heuser, Beatrice, *The Evolution of Strategy: Thinking War from Antiquity to the Present* (Cambridge University Press, 2010).

Hoffman, Frank, 'Hybrid Warfare and Challenges', *Joint Forces Quarterly* 52 (2009), 34–9.

Horgan, John, 'Cross-check: Obama's Choice for Warrior in Chief, Gen. James Mattis, Calls Iraq Invasion "the Dumbest Thing We Ever Did"', *Scientific American*, 27 July 2010, http://blogs.scientificamerican.com/cross-check/2010/07/27/obamas-choice-for-warrior-in-chief-gen-james-mattis-calls-iraq-invasion-the-dumbest-thing-we-ever-did/ (10 May 2012).

Horne, Alistair, *To Lose a Battle: France 1940* (London: Penguin, 2007).

Horne, Alistair, *A Savage War of Peace: Algeria 1954–1962* (New York: New York Review Books Classics, 2006).

House of Commons Public Administration Select Committee, Who Does UK National Strategy? First Report of Session 2010–11, Ordered by the House of Commons to be printed 12 October 2010, www.publications.parliament.uk/pa/cm201011/cmselect/cmpubadm/435/435.pdf (13 November 2012).

House of Commons Public Administration Select Committee, *Who Does UK National Strategy? Further Report*, Sixth Report of Session 2010–11, Ordered by the House of Commons to be printed 25 January 2011, www.publications.parliament.uk/pa/cm201011/cmselect/cmpubadm/713/713.pdf (13 November 2012).

Huffington Post, 'Raid In Somalia: U.S. Navy SEALs Free Hostages Held Since October', January 2012, www.huffingtonpost.com/2012/01/25/raid-in-omalia_n_1230062.html (2 February 2012).

Huntington, Samuel, *The Soldier and the State: The Theory and Politics of Civil–Military Relations* (Cambridge, MA: Belknap Press, 1964).

iCasualties: Operation Iraqi Freedom and Operation Enduring Freedom Casualties, 2009, http://icasualties.org/ (26 September 2012).

International Human Rights and Conflict Resolution Clinic (Stanford Law School) and Global Justice Clinic (NYU School of Law), *Living under Drones: Death, Injury, and Trauma to Civilians from US Drone Practices in Pakistan* (September 2012), http://livingunderdrones.org/download-report/ (3 October 2012).

International Institute for Strategic Studies, *The Military Balance* (London: Routledge, 2012).

Jenkins, Roy, *Churchill: A Biography* (London: Pan Books, 2002).

Johnson, David E., M. Wade Markel and Brian Shannon, *The 2008 Battle of Sadr City*, prepared for the United States Army (Santa Monica, CA: RAND Corporation, 2011).

Joy, Bill, 'Why the Future Doesn't Need Us', *Wired* 8(04) (April 2000), www.wired.com/wired/archive/8.04/joy.html?pg=1&topic=&topic_set= (13 September 2012).

Kalberg, Stephen, 'Max Weber's Types of Rationality: Cornerstones for the Analysis of Rationalization Processes in History', *The American Journal of Sociology* 85 (1980), 1145–79.

Kaplan, Robert D., 'Man versus Afghanistan', *The Atlantic*, April 2010, www.theatlantic.com/magazine/archive/2010/04/man-versus-afghanistan/307983/ (2 October 2012).

Keegan, John, *The Face of Battle: A Study of Agincourt, Waterloo and the Somme* (London: Pimlico, 1991).

Keegan, John, *The First World War* (London: Vintage, 2000).

Kendall, Frank, *Testimony Before the House Armed Services Committee*, Witness Statement of Hon. Frank Kendall, Under Secretary of Defense Acquisition, Technology & Logistics, 28 January 2014, US Congress, House of Representatives, Armed Services Committee.

King, Anthony, *The Transformation of Europe's Armed Forces: From the Rhine to Afghanistan* (Cambridge University Press, 2011).

Kitchen, Martin, 'Ludendorff and Germany's Defeat', in Hugh Cecil and Peter H. Liddle (eds.), *Facing Armageddon: The First World War Experienced* (London: Leo Cooper, 1996), 51–66.

Koch, Richard, *Strategy: How to Create, Pursue and Deliver a Winning Strategy* (Harlow: Prentice Hall, 2011).

Krepinevich, Andrew F., *Defense Investment Strategies in an Uncertain World* (Washington, DC: Center for Strategic and Budgetary Assessment, 2008).

Krepinevich, Andrew F. and Barry D. Watts, *Regaining Strategic Competence* (Washington, DC: Center for Strategic and Budgetary Assessment, 2009).

Kwak, Young Hoon and Brian M. Smith, 'Managing Risks in Mega Defense Acquisition Projects: Performance, Policy, and Opportunities', *International Journal of Project Management* 27 (2009), 812–20.

Lave, Jean and Etienne Wenger, *Situated Learning: Legitimate Peripheral Participation* (Cambridge University Press, 1991).

Lawrence, T. E., *Seven Pillars of Wisdom* (Ware: Wordsworth Editions, 1997).

Leavy, Brian, 'Design Thinking: A New Mental Model for Value Innovation', *Strategy and Leadership* 38 (2010).

Liddell Hart, Basil H. , *Strategy: The Indirect Approach* (New Delhi: Natraj, 2003).

Liddell Hart, Basil H., 'War, Limited', *Harper's Magazine* 192 (March 1946).

Lostumbo, Michael J. *et al.*, *Overseas Basing of U.S. Military Forces: An Assessment of Relative Costs* (Washington, DC: RAND, 2013).

Luhmann, Niklas, *Risk: A Sociological Theory*, transl. Rhodes Barrett (New York: Walter de Gruyter, 1993).

Luhmann, Niklas, *Social Systems* (Stanford University Press, 1995).

Mackay, Andrew and Steve Tatham, *Behavioural Conflict: Why Understanding People and Their Motives Will Prove Decisive in Future Conflict* (Saffron Walden: Military Studies Press, 2011).

Mackinlay, John, 'After 2015: The Next Security Era for Britain', *Prism* 3 (2012), 51–61.

Macleod, Ken, 'Politics and Science Fiction', in Edward James and Farah Mendlesohn, *The Cambridge Companion to Science Fiction* (Cambridge University Press, 2003), 230–41.

March, James, and Johan P. Olsen, *Rediscovering Institutions: The Organizational Basis of Politics* (New York: The Free Press, 1989).

Marines.mil – the Official Homepage of the United States Marine Corps, www.hqmc.marines.mil/News/NewsArticleDisplay/tabid/3488/Article/78942/the-air-sea-battle-concept-summary.aspx (27 September 2012).

Markides, Constantinos C., *Game-Changing Strategies: How to Create New Market Space in Established Industries by Breaking the Rules* (San Francisco, CA: Jossey-Bass, 2008).

Marshall, Andrew W., *Some Thoughts on Military Revolutions*, Second Version, Office of the Secretary of Defense, Director of Net Assessment, Washington, DC (1993).

Marshall, Andrew, *Strategy for Competing With the Soviets in the Military Sector of the Continuing Political-Military Competition*, Department of Defense Office of Net Assessment, 1976, mimeo, http://goodbadstrategy.com/wp-content/downloads/StrategyforCompetingwithUSSR.pdf (20 January 2012).

Martin, Roger, *The Design of Business: Why Design Thinking is the Next Competitive Advantage* (Boston, MA: Harvard Business Press, 2009).

Maslow, Abraham H., *The Psychology of Science* (New York: Joanna Cutler Books, 1966).

McGrath, John J., *The Other End of the Spear: The Tooth-to-Tail Ratio (T3R) in Modern Military Operations* (Fort Leavenworth, KS: Combat Studies Institute Press, 2007).

Meadows, Donella, *Thinking in Systems* (London: Routledge, 2009).

Meadows, Donella, Dennis Meadows, Jørgen Randers and William W. Behrens III, *Limits to Growth: A Report for the Club of Rome's Project on the Predicament of Mankind* (London: Signet, 1972).

Metz, Steven, *Armed Conflict in the 21st Century: The Information Revolution and Post-Modern Warfare* (Carlisle, PA: Strategic Studies Institute, US Army War College, 2000).

Mills, Greg 'Coke Isn't It: Changing a Culture and Image of Violence', in David Richards and Greg Mills (eds.), *Victory Among People: Lessons from Counter Insurgency and Stabilising Fragile States* (London: RUSI, 2011), 383–412.

Ministry of Defence, *Global Strategic Trends: Out to 2040*, Strategic Trends Programme, 4th edition, 2010, www.mod.uk/NR/rdonlyres/6AAFA4FA-C1D3-4343-B46F-05EE80314382/0/GST4_v9_Feb10.pdf (28 August 2012).

Moltke, Helmuth von, '1869 Instructions for Large Unit Commanders', in Daniel J. Hughes (ed.), *Moltke on the Art of War: Selected Writings* (Novato, CA: Presido Press, 1993).

Mölling, Christian, *Europe Without Defence*, SWP Comments 38, November 2011, Stiftung Wissenschaft und Politik, Berlin, www.swp-berlin.org/fileadmin/contents/products/comments/2011C38_mlg_ks.pdf (13 November 2012).

Nagl, John, 'The Age of Unsatisfying Wars', *New York Times*, 6 June 2012, www.nytimes.com/2012/06/07/opinion/the-age-of-unsatisfying-wars.html?_r=1 (7 June 2012).

NATO, *Deterrence and Defence Posture Review*, 20 May 2012, www.nato.int/cps/en/natolive/official_texts_87597.htm?mode=pressrelease (6 November 2012).

NATO, *Strategic Concept For the Defence and Security of The Members of the North Atlantic Treaty Organisation*, Adopted by Heads of State and Government in Lisbon, November 2010, www.nato.int/lisbon2010/strategic-concept-2010-eng.pdf (3 October 2012).

NATO, *Summit Declaration on Defence Capabilities: Toward NATO Forces 2020*, 20 May 2012, www.nato.int/cps/en/natolive/official_texts_87594.htm?mode=pressrelease (12 December 2012).

Nevin, Thomas, 'Ernest Jünger: German Stormtrooper Chronicler', in Hugh Cecil and Peter H. Liddle (eds.), *Facing Armageddon* (London: Leo Cooper, 1996), 269–77.

New York Times, 'Enemy Lurks in Briefings on Afghan War – PowerPoint', 27 April 2010, www.nytimes.com/2010/04/27/world/27powerpoint.html (10 May 2012).

New York Times, 'Globalizaton Creates a New Worry: Enemy Convergence', 30 May 2013.

New York Times, 'Trading Program Ran Amok, With No "Off" Switch', Dealbook, newyorktimes.com, 3 August 2012, http://dealbook.nytimes.com/2012/08/03/trading-program-ran-amok-with-no-off-switch/ (13 September 2012).

New York Times, 'US Army Hones Antiterror Strategy for Africa, in Kansas', 18 October 2013.

New York Times, 'US Defense Chief Tours 9/11 Memorial', 7 September 2011.

Nordström, Kjell A. and Jonas Ridderstråle, *Funky Business: Talent Makes Capital Dance* (London: Pearson, 2000).

Obama, Barack, *Press Conference with President Obama and President Park of the Republic of Korea*, Blue House, Seoul, Republic of Korea, 25 April 2014, The White House, www.whitehouse.gov/the-press-office/2014/04/25/press-conference-president-obama-and-president-park-republic-korea (28 April 2014).

Obama, Barack, *Remarks by the President on the Defense Strategic Review*, The White House. 2012, www.whitehouse.gov/the-press-office/2012/01/05/remarks-president-defense-strategic-review (20 January 2012).

Obama, Barack, *Remarks by the President and First Lady on the End of the War in Iraq*, The White House, 2011 www.whitehouse.gov/the-press-office/2011/12/14/remarks-president-and-first-lady-end-war-iraq, (30 September 2012).

Obama, Barack, *Remarks by the President at the National Defense University*, National Defense University, Fort McNair, Washington, DC, 23 May 2013, www.whitehouse.gov/the-press-office/2013/05/23/remarks-president-national-defense-university (19 November 2013).

The Observer, 'Obama's Secret Kill List: The Disposition Matrix', 14 July 2013.

O'Hanlon, Michael, *Healing the Wounded Giant: Maintaining Military Preeminence While Cutting the Defense Budget* (Washington, DC: The Brookings Institution, 2013).

O'Hanlon, Michael, *The Science of War: Defense Budgeting, Military Technology, Logistics, and Combat Outcomes* (Princeton University Press, 2009).

Ostroff, Frank, 'Change Management in Government', *Harvard Business Review* May (2006), 141–7.

Owens, Bill (with Ed Offley), *Lifting the Fog of War* (New York: Farrar, Straus and Giroux, 2000).

Panetta, Leon, *Defending the Nation from Cyber Attack (Business Executives for National Security)*, as Delivered by Secretary of Defense Leon E. Panetta, New York, 11 October 2012, www.defense.gov/speeches/speech.aspx?speechid=1728, (30 October 2012).

Paparone, Christopher R. and George Reed, 'The Reflective Military Practitioner: How Military Professionals Think in Action', *Military Review* 88(2) (March–April 2008), http://usacac.army.mil/CAC2/MilitaryReview/Archives/English/MilitaryReview_20080430_art011.pdf (7 June 2012).

PBS, 'Kill/Capture', *Frontline*, Transcript, www.pbs.org/wgbh/pages/frontline/afghanistan-pakistan/kill-capture/transcript/ (2 October 2012).

Petraeus, David Howell, *The American Military and the Lessons of Vietnam: A Study of Military Influence and the Use of Force in the Post-Vietnam Era*, a dissertation presented to the Faculty of Princeton University in Candidacy for the Degree of Doctor of Philosophy, Woodrow Wilson School of Public and International Affairs, mimeo, October 1987.

Porter, Wayne and Mark Mykleby, *A National Strategic Narrative*, Woodrow Wilson Center, 2011, www.wilsoncenter.org/sites/default/files/A%20National%20Strategic%20Narrative.pdf (30 September 2012).

Rasmussen, Mikkel Vedby, *The Risk Society at War* (Cambridge University Press, 2006).

Rasmussen, Mikkel Vedby and Henrik Ø. Breitenbauch, *Denmark's Need for Fighter Aircraft* (Copenhagen: Danish Institute for Military Studies, 2007).

Rasmussen, Mikkel Vedby and Jon Rhabek-Clemmensen, *Trends in NATO Defence Budgets*, mimeo, (Copenhagen: Centre for Military Studies, 2014).

Richards, David, 'Twenty-first Century Armed Forces: Agile, Useable, Relevant', 23 June 2009, RUSI, Whitehall, London, www.rusi.org/events/ref:E496B737B57852/info:public/infoID:E4A4253226F582/#.UoVSMRajPq8 (14 November 2013).

Ricks, Thomas E., *The Gamble: General Petraeus and the Untold Story of the American Surge in Iraq* (London: Penguin, 2010).

Ricks, Thomas E., *The Generals: American Military Command from World War II to Today* (New York: Penguin, 2012).

The Rio Times, 'Military Forces Remain in Complexo do Alemão and Penha in Rio's Zona Norte (North Zone)', 15 May 2012, http://riotimesonline.com/brazil-news/rio-politics/military-forces-remain-in-complexo-do-alemao/ (7 June 2012).

The Rio Times, 'Occupation of Zona Norte Favelas', 28 December 2010, http://riotimesonline.com/brazil-news/rio-politics/occupation-of-zona-norte-favelas/ (7 June 2012).

Roco, Mihail C. and William Sims Bainbridge (eds.), *Converging Technologies for Improving Human Performance: Nanotechnology, Biotechnology, Information Technology and Cognitive Science*, National Science Foundation report (Dordrecht: Kluwer Academic, 2003), www.wtec.org/ConvergingTechnologies/Report/NBIC_report.pdf (27 September 2012).

Rotmann, Philipp, *Built on Shaky Ground: The Comprehensive Approach in Practice*, Research Paper, No 63, December 2010 (Rome: NATO Defence College, 2010).

Rumlet, Richard, *Good Strategy/Bad Strategy: The Difference and Why It Matters* (London: Profile Books, 2011).

Sanger, David E., *Confront and Conceal: Obama's Secret Wars and Surprising Use of American Power* (New York: Random House, 2012).

Santos, Juan Manuel, 'Foreword', in David Richards and Greg Mills (eds.), *Victory Among People: Lessons from Counter Insurgency and Stabilising Fragile States* (London: RUSI, 2011).

Schelling, Thomas, *Arms and Influence* (New Haven, CT: Yale University Press, 1966).

Schindler, John R., 'How to Win Cold War 2.0', *Politico*, 25 March 2014, www.politico.com/magazine/story/2014/03/new-cold-war-russia-104954.html#.U1319cYbZ8D (28 April 2014).

Shapiro, Jacob N., *The Terrorist's Dilemma. Managing Violent Covert Organizations* (Princeton University Press, 2013).

Sheehan, James, *The Monopoly of Violence: Why Europeans Hate Going to War* (London: Faber and Faber, 2008).

Shimko, Keith L., *The Iraq Wars and America's Military Revolution* (Cambridge University Press, 2010).

Simonis, Frank and Steven Schilthuizen, *Nanotechnology: Innovation Opportunities for Tomorrow's Defence*, 2006, www.futuretechnologycenter.eu/downloads/nanobook.pdf (6 November 2012).

Singer, Peter W., *Wired for War: The Robotics Revolution and Conflict in the 21st Century* (London: Penguin, 2009).

Skinner, B. F., *Beyond Freedom and Dignity* (London: Penguin, 1971).

Slack, Nigel and Michael Lewis, *Operations Strategy*, 3rd edition (Harlow: Prentice Hall, 2011).

Smith, Adam, *An Inquiry into the Nature and Causes of the Wealth of Nations*, (Indianapolis, IN: Liberty Classics, 1981).

Smith, Rupert, *The Utility of Force: The Art of War in the Modern World* (London: Vintage, 2008).

Stirrup, Jock, *Annual Chief of Defence Staff Lecture*, RUSI, Whitehall, London, 3 December 2009, www.rusi.org/events/past/ref:E4B184DB05C4E3/ (13 November 2012).

Strachan, Hew, *The Direction of War: Contemporary Strategy in Historical Perspective* (Cambridge University Press, 2013).

Strategic Landpower Task Force, *Strategic Landpower: Winning the Clash of Wills*, May 2013, www.ausa.org/news/2013/Documents/Strategic%20Landpower%20White%20Paper%20May%202013.pdf (13 November 2013).

Struwe, Lars Bangert, 'Private Security Companies (PSCs) as a Piracy Countermeasure', *Studies in Conflict and Terrorism* 35 (2012), 588–96.

Sun Tzu, *The Art of War*, translated and with an introduction by Samuel B. Griffith (Oxford University Press, 1963).

Swidler, Ann, 'The Concept of Rationality in the Work of Max Weber', *Sociological Inquiry* 43 (1973), 35–42.

Swidler, Anne, 'Culture in Action: Symbols and Strategies', *American Sociological Review* 51 (1986), 273–86.

Taptiklis, Theodore, *Unmanaging: Opening Up the Organization to Its Own Unspoken Knowledge* (London: Palgrave Macmillan, 2008).

Taylor, Maxwell D., *The Uncertain Trumpet* (London: Atlantic Books, 1959).

The Telegraph, 'Bomber Command Veterans Boycotting "Insulting" Award', 17 May 2013, www.telegraph.co.uk/history/britain-at-war/10064299/Bomber-Command-veterans-boycotting-insulting-award.html (19 November 2013).

Terriff, Terry, Theo Farrell and Frans Osinga, *A Transformation Gap: American Innovations and European Military Change* (Stanford University Press, 2010).

Thaler, Richard and Cass Sunstein, *Nudge* (London: Penguin, 2009).

Thomas, Jim, *Department of Defense Investment in Technology and Capability to Meet Emerging Threats*, Testimony Before the US House of Representatives, (Washington, DC: Center for Strategic and Budgetary Assessment, 2011).

Tol, Jan van, Mark Gunzinger, Andrew F. Krepinevich and Jim Thomas, *AirSea Battle: A Point-of-Departure Operational Concept* (Washington, DC: Center for Strategic and Budgetary Assessment, 2010).

Trinquier, Roger, *Modern Warfare: A French View of Counterinsurgency* (Westport, CT: Praeger, 2006).

Ucko, David H., *The New Counterinsurgency Era: Transforming the U.S. Military for Modern Wars* (Washington, DC: Georgetown University Press, 2009).

Ucko, David H. and Robert Egnell, *Counterinsurgency in Crisis: Britain and the Challenges of Modern Warfare* (New York: Columbia University Press, 2013).

UK Government, *Securing Britain in an Age of Uncertainty: The Strategic Defence and Security Review*, Presented to Parliament by the Prime Minister by Command of Her Majesty, October 2010, http://webarchive.nationalarchives. gov.uk/2012101500000/http: www.direct.gov.uk/prod_conum_dg/groups/dg_ digitalassets/@dg/@en/documents/digitalasset/dg_191634. pdf (29 November 2014).

UK Supreme Court, *Smith and others (Appellants)* v. *The Ministry of Defence (Respondent); Ellis (Respondent)* v. *The Ministry of Defence (Appellant); Allbutt and others (Respondents)* v. *The Ministry of Defence (Appellant) [2013] UKSC 41*, 19 June 2013, www.supremecourt.gov.uk/decided-cases/ docs/UKSC_2012_0249_PressSummary.pdf (27 November 2013).

United Nations Population Fund, www.un.org/esa/population/publications/ wpp2008/pressrelease.pdf (25 May 2012).

US Army, M 3–06.11 (FM 90-10-1), *Combined Arms Operations in Urban Terrain*, Headquarters No. 3–06.11. Headquarters, Department of the Army, Washington, DC, 28 February 2002, 1–1.

US Army, *Counterinsurgency*, FM 3–24, MCWP 3–33.5, December 2006, www.fas.org/irp/doddir/army/fm3-24.pdf (7 June 2012).

US Army, 'TRADOC: Strategic Landpower Concept to Change Doctrine', 16 January 2014, www.army.mil/article/118432/TRADOC__Strategic_Landpower_con- cept_to_change_doctrine/ (28 January 2014).

US Department of Defense, *National Defense Budget Estimates for FY 2011*, Office for the Undersecretary of Defense (comptroller), March 2011, http://comptroller.defense.gov/Portals/45/Documents/defbudget/fy2011/FY11_ Green_Book.pdf (26 April 2012), Table 6-13.

US Department of Defense, *National Defense Budget Estimates for FY 2014*, Department of Defense, Office of the Under Secretary of Defense (Comptroller), May 2013, http://comptroller.defense.gov/Portals/45/Documents/defbudget/fy2014/FY14_Green_Book.pdf (26 November 2013).

US Department of Defense, *News Transcript: Secretary Rumsfeld Town Hall Meeting in Kuwait*, 8 December 2004, www.defense.gov/transcripts/transcript.aspx?transcriptid=1980 (25 September 2012).

US Department of Defense, *Quadrennial Defense Review Report*, May 1997 (Washington, DC: Department of Defense, 1997).

US Department of Defense, *Quadrennial Defense Review Report*, 30 September 2001 (Washington, DC: Department of Defense, 2001).

US Department of Defense, *Quadrennial Defence Review Report*, February 2010, www.defense.gov/qdr/images/QDR_as_of_12Feb10_1000.pdf (25 September 2012).

US Department of Defense, *Quadrennial Defence Review 2014*, March 2014, www.defense.gov/pubs/2014_Quadrennial_Defense_Review.pdf (28 March 2014).

US Department of Defense, *Sustaining US Global Leadership: Priorities for 21st Century Defence*, January 2012, www.defense.gov/news/Defense_Strategic_Guidance.pdf (20 January 2012).

US Joint Chiefs of Staff, *Chairman's Direction to the Joint Force*, 6 February 2012, http://usacac.army.mil/cac2/repository/020312135111_CJCS_Strategic_Direction_to_the_Joint_Force_6_Feb_2012.pdf (8 November 2012)

US Joint Chiefs of Staff, *Decade of War, Volume I: Enduring Lessons from the Past Decade of Operations*, Joint and Coalition Operational Analysis, a division of the Joint Staff J7.

US Joint Chiefs of Staff, *Joint Vision 2010* (Washington, DC: Chairman of the Joint Chiefs of Staff, 1997) www.dtic.mil/jv2010/jv2010.pdf.

US Joint Forces Command, *Capstone Concept for Joint Operations*, 15 January 2009, www.jfcom.mil/newslink/storyarchive/2009/CCJO_2009.pdf (8 November 2012).

US National Security Council, 'A Report to the National Security Council by the Executive Secretary on United States Objectives and Programs for National Security', 14 April 1950, reprinted in Dennis Merrill (ed.), *Documentary History of the Truman Presidency, Volume 7: The Ideological Foundations of the Cold War – the 'Long Telegram', the Clifford Report, and NSC 68* (University Publications of America, 1996).

US Senate, Hearing to Consider the Nomination of General Martin E. Dempsey, USA, For Reappointment to the Grade of General and to Be Chief of Staff,

United States Army, www.armed-services.senate.gov/Transcripts/2011/03%20 March/11-08%20-%203-3-11.pdf (15 November 2012).

Vandergriff, Donald, *The Path to Victory: America's Army and the Revolution in Human Affairs* (Novato, CA: Presidio Press, 2002).

Vego, Milan, 'Is the Conduct of War a Business?', *Joint Forces Quarterly* 59 (4th Quarter 2010), 57–65.

Vinge, Vernor, *The Coming Technological Singularity: How To Survive in the Post-Human Era*, 1993, http://mindstalk.net/vinge/vinge-sing.html (13 September 2012).

Walker, Jeremy and Melinda Cooper, 'Genealogies of Resilience: From Systems Ecology to the Political Economy of Crisis Adaptation', *Security Dialogue* 42 (2011), 143–60.

Walt, Stephen M., 'Is America Addicted to War?', *Foreign Policy*, 4 April 2011, www.foreignpolicy.com/articles/2011/04/04/is_america_addicted_to_ war?page=full (13 November 2012).

Warden, John, 'Air Theory for the Twenty-first Century', in Barry R. Schneider and Lawrence E. Grinter (eds.), *Battlefield of the Future: 21st Century Warfare Issues*, The Air War College Studies in National Security (Maxwell Air Force Base, AL: Air War College, 1995).

Warren, Robert, 'Situating the City and September 11th: Military Urban Doctrine, Pop-Up Armies and Spatial Chess', *International Journal of Urban and Regional Research*, 26(3) (September 2002), 614–19.

Washington Post, 'An Army Takeover Quells Violence in Mexico', 21 April 2009.

Wasserbly, Daniel, 'Obama Looks to Leaner Military, As-Pac Focus', *Jane's Defence Weekly* 49 (2012), 4–5.

Watts, Barry D., *The Maturing Revolution in Military Affairs* (Washington, DC: Center for Strategic and Budgetary Assessments, 2011).

Weber, Max, *From Max Weber: Essays in Sociology*, H. H. Gerth and C. Wright Mills (eds.) (London: Routledge, 1997).

Weiner, Sharon K., 'Organisational Interests versus Battlefield Needs: The U.S. Military and Mine-Resistant Ambush Protected Vehicles in Iraq', *Polity*, 42(4) (October 2010), 461–82.

Windrow, Martin, *The Last Valley* (London: Cassell, 2005).

Wired.com, 'How to Defeat the Air Force's Powerful Stealth Fighter', 30 July 2012, www.wired.com/dangerroom/2012/07/f-22-germans/ (13 September 2012).

Index